KEYS TO
Science Success

Assig 1 A: Read ch 2 + 3
B: BB Discussion
   4-5 sentence News
   Report
✓ C. p. 55-57; 59-60; 61

Janet R. Katz

Carol Carter   Joyce Bishop   Sarah Lyman Kravits

Prentice Hall
Upper Saddle River, New Jersey 07458

Library of Congress Cataloging-in-Publication Data

Keys to science success / Janet R. Katz . . . [et al.].
   p.   cm.
   Includes bibliographical references and index.
   ISBN 0-13-013305-1
   1. Science—Vocational guidance.   I. Katz, Janet R.
Q147.K49 2000
502.3—dc21   99-30050
                                                                CIP

Publisher: *Carol Carter*
Acquisitions Editor: *Sue Bierman*
Managing Editor: *Mary Carnis*
In-House Liaison: *Glenn Johnston*
Production: *Holcomb Hathaway, Inc.*
Director of Manufacturing and Production: *Bruce Johnson*
Manufacturing Buyer: *Marc Bove*
Cover designer: *Bruce Kenselaar*
Cover illustration: *Paul Gourhan*
Editorial Assistant: *Michelle M. Williams*
Marketing Manager: *Jeff McIlroy*
Marketing Assistant: *Barbara Rosenberg*

Copyright © 2000 by Prentice-Hall, Inc.
Upper Saddle River, New Jersey 07458

All rights reserved. No part of this publication may be reproduced, stored in a retrieval system, or transmitted in any form or by any means, electronic, mechanical, photocopy, recording, or otherwise, without the prior written permission of the publisher.

Printed in the United States of America

10 9 8 7 6 5 4 3 2 1

ISBN 0-13-013305-1

Prentice-Hall International (UK) Limited, London
Prentice-Hall of Australia Pty. Limited, Sydney
Prentice-Hall Canada Inc., Toronto
Prentice-Hall Hispanoamericana, S.A., Mexico
Prentice-Hall of India Private Limited, New Delhi
Prentice-Hall of Japan, Inc., Tokyo
Prentice-Hall Pte. Ltd., Singapore
Editora Prentice-Hall do Brasil, Ltda., Rio de Janeiro

Brief Contents

Preface xiii
About the Authors xix

1 **RESEARCHING YOUR SCIENCE EDUCATION**   1
Collecting the Basic Data

2 **DISCOVERING SCIENCE**   25
Exploring Your Options

3 **SELF-AWARENESS**   51
Knowing Who You Are and How You Learn

4 **GOAL SETTING AND TIME MANAGEMENT**   81
Mapping Your Course

5 **SCIENTIFIC INQUIRY**   107
Critical and Creative Thinking

6 **READING AND STUDYING**   137
Maximizing Written Resources

7 **NOTE-TAKING AND WRITING**   167
Harnessing the Power of Words and Ideas

8 **LISTENING, MEMORY, AND TEST TAKING**   197
Taking In, Retaining, and Demonstrating Knowledge

9 **WORKING IN THE LABORATORY** 223
Safe Science in Action

10 **RELATING TO OTHERS** 237
Appreciating Your Diverse World

11 **MANAGING CAREER AND MONEY** 267
Reality Resources

12 **MOVING AHEAD** 293
Building a Smart Future

Endnotes 321

Index 325

Contents

Preface xiii
About the Authors xix

1 RESEARCHING YOUR SCIENCE EDUCATION 1
Collecting the Basic Data

WHY DO YOU NEED TO STUDY A VARIETY OF SCIENCES? 2
WHO IS PURSUING SCIENCE EDUCATION TODAY? 3
    The Diverse Student Body 3
    Women in Science 5
HOW DO SCIENCE KNOWLEDGE AND SKILLS PROMOTE SUCCESS? 5
WHAT RESOURCES ARE AVAILABLE AT YOUR SCHOOL? 9
    People 9
    Student Services 13
    Organizations 13
    College Catalogs and Student Handbooks 14
WHAT IS YOUR ROLE IN A DIVERSE WORLD? 16
    Diversity Is Real in Science 16
    Ethnocentrism 17
    Diversity and Teamwork 17
REAL WORLD PERSPECTIVE 18
    Living Your Role 19
APPLICATIONS 20

2 DISCOVERING SCIENCE 25
Exploring Your Options

HOW CAN YOU FIND WHAT FASCINATES YOU? 26

WHAT SKILLS DO YOU NEED TO DEVELOP TO SUCCEED IN SCIENCE?   27
WHAT SKILLS DO YOU ALREADY HAVE TO SUCCEED IN SCIENCE?   28
    Science Ability   28
    Interest in Science   29
WHAT ARE THE FIELDS OF STUDY AND CAREERS IN SCIENCE?   29
    Biological Science   30
    Health Science   32
    Physical Science   37
    Mathematics   39
    Computer Science   39
    Engineering Science   41
    Alternative Careers in Science   42
WHAT ARE SOME OF THE BARRIERS WOMEN FACE IN THE TECHNICAL SCIENCES?   42
    The Barriers   42
    The Solutions   43
REAL WORLD PERSPECTIVE   43
    Men in Engineering Today   44
HOW CAN YOU START THINKING ABOUT CHOOSING A SPECIFIC SCIENCE MAJOR?   44
    Exploring Potential Majors   44
    Changing Majors   45
    Linking Majors to Career Areas   45
APPLICATIONS   47

3 SELF-AWARENESS   51
Knowing Who You Are and How You Learn

HOW WELL DID YOUR SCHOOL PREPARE YOU FOR COLLEGE SCIENCE?   52
IS THERE ONE BEST WAY TO LEARN?   53
HOW CAN YOU DISCOVER YOUR LEARNING STYLES?   53
    Learning Styles Inventory   53
    Multiple Intelligences Theory   58
    Personality Spectrum   60
WHAT ARE THE BENEFITS OF KNOWING YOUR LEARNING STYLES?   62
    Study Benefits   62
    General Benefits   67
HOW DO YOU EXPLORE WHO YOU ARE?   68
REAL WORLD PERSPECTIVE   69
    Self-Perception   69
    Interests   70
    Habits   71
    Abilities   72
APPLICATIONS   75

CONTENTS   vii

4 GOAL SETTING AND TIME MANAGEMENT   81
Mapping Your Course

WHAT DEFINES YOUR VALUES?   82
    Choosing and Evaluating Values   82
    How Values Relate to Goals   83
HOW DO YOU SET AND ACHIEVE GOALS?   83
    Identifying Your "Personal Mission Statement"   83
    Placing Goals in Time   84
    Linking Goals With Values   87
WHAT ARE YOUR PRIORITIES?   89
HOW CAN YOU MANAGE YOUR TIME?   89
REAL WORLD PERSPECTIVE   91
    Building a Schedule   91
REAL WORLD PERSPECTIVE   95
    Time Management Strategies   96
WHY IS PROCRASTINATION A PROBLEM?   98
    Strategies to Fight Procrastination   98
APPLICATIONS   100

5 SCIENTIFIC INQUIRY   107
Critical and Creative Thinking

WHAT IS CRITICAL THINKING?   108
    Critical Thinking Is a Skill   108
    A Critical-Thinking Response to a Statement   110
    The Value of Critical Thinking   111
HOW IS CRITICAL THINKING CRITICAL IN SCIENCE?   111
HOW DOES YOUR MIND WORK?   112
    Mind Actions: The Thinktrix   113
    How Mind Actions Build Thinking Processes   116
HOW DOES CRITICAL THINKING HELP YOU SOLVE PROBLEMS
  AND MAKE DECISIONS?   117
    Problem Solving and Inquiry   117
    Decision Making   119
WHY SHIFT YOUR PERSPECTIVE?   123
REAL WORLD PERSPECTIVE   124
WHY PLAN STRATEGICALLY?   125
HOW CAN YOU DEVELOP CREATIVITY IN SCIENCE?   127
    Characteristics of Creative People in Science   128
    Brainstorming Toward a Creative Answer   130

Creativity and Critical Thinking 131
APPLICATIONS 132

6 READING AND STUDYING 137
Maximizing Written Resources

WHAT ARE SOME CHALLENGES IN SCIENCE READINGS? 138
    Dealing With Reading Overload 138
    Working Through Difficult Science Texts 138
    Managing Distractions 140
REAL WORLD PERSPECTIVE 141
    Building Comprehension and Speed 142
WHAT KIND OF READING WILL YOU DO IN THE SCIENCES? 143
WHY DEFINE YOUR PURPOSE FOR READING? 144
    Purpose Determines Reading Strategy 145
    Purpose Determines Pace 145
HOW CAN PQ3R HELP YOU STUDY READING MATERIALS? 146
    Preview-Question-Read-Recite-Review (PQ3R) 147
    Preview 147
    Question 147
    Read 148
    Recite 150
    Review 150
HOW CAN YOU READ CRITICALLY? 151
    Use PQ3R to "Taste" Reading Material 151
    Ask Questions Based on the Mind Actions 153
    Shift Your Perspective 154
    Seek Understanding 155
WHAT RESOURCES DOES YOUR LIBRARY OFFER? 155
    General Reference Works 155
APPLICATIONS 158

7 NOTE-TAKING AND WRITING 167
Harnessing the Power of Words and Ideas

HOW DOES TAKING NOTES HELP YOU? 168
    Recording Information in Class 168
    Make Notes a Valuable After-Class Reference 170
WHICH NOTE-TAKING SYSTEM SHOULD YOU USE? 170
    Taking Notes in Outline Form 170
    Using the Cornell Note-Taking System 172
    Creating a Think Link 172
HOW CAN YOU WRITE FASTER WHEN TAKING NOTES? 173

WHY DOES GOOD WRITING MATTER IN SCIENCE? 175
WHAT ARE THE ELEMENTS OF EFFECTIVE WRITING? 176
    Writing Purpose 176
    Knowing Your Audience 177
WHAT IS THE WRITING PROCESS IN SCIENCE? 178
    Planning 178
    Drafting 182
    Revising 184
    Editing 187
    The Research Format 187
APPLICATIONS 191

8 LISTENING, MEMORY, AND TEST TAKING 197
Taking In, Retaining, and Demonstrating Knowledge

HOW CAN YOU BECOME A BETTER LISTENER? 198
    Manage Listening Challenges 198
HOW DOES MEMORY WORK? 200
HOW CAN YOU IMPROVE YOUR MEMORY? 201
    Memory Improvement Strategies 201
    Mnemonic Devices 203
HOW CAN TAPE RECORDERS HELP YOU LISTEN, REMEMBER, AND STUDY? 204
HOW CAN PREPARATION HELP IMPROVE TEST SCORES? 206
    Identify Test Type and Material Covered 206
    Use Specific Study Skills 207
    Prepare Physically 208
    Conquer Test Anxiety 208
WHAT STRATEGIES CAN HELP YOU SUCCEED ON WRITTEN TESTS? 209
    Write Down Key Facts 209
    Begin With an Overview of the Exam 209
    Know the Ground Rules 210
    Use Critical Thinking to Avoid Errors 210
    Master Different Types of Test Questions 211
HOW CAN YOU LEARN FROM TEST MISTAKES? 215
APPLICATIONS 217

9 WORKING IN THE LABORATORY 223
Safe Science in Action

WHAT IS INQUIRY-BASED RESEARCH? 224
    Qualities of a Researcher 225
    Levels of Inquiry 225
WHAT EQUIPMENT WILL YOU NEED IN THE LAB? 226

HOW DO COURSE CREDIT HOURS TRANSLATE INTO LAB TIME? 226
WHAT CAN YOU DO TO BOOST YOUR CHANCES FOR SUCCESS? 227
    Preparation   227
    Attendance   227
    Writing   228
    Curiosity   228
WHY IS IT IMPORTANT TO PRACTICE SAFE LAB SCIENCE? 228
WHAT SKILLS DOES LAB WORK GIVE YOU? 230
    Communication Skills   230
    Teamwork Skills   230
    Critical-Thinking Skills   231
    Technical Skills   231
    A Summary for Lab Success   231
APPLICATIONS   *233*

10 RELATING TO OTHERS 237
Appreciating Your Diverse World

HOW CAN YOU UNDERSTAND AND ACCEPT OTHERS? 238
    Diversity in Your World   238
    The Positive Effects of Diversity   239
    Barriers to Understanding   240
REAL WORLD PERSPECTIVE 245
    Accepting and Dealing With Differences   246
HOW CAN YOU EXPRESS YOURSELF EFFECTIVELY? 248
    Adjusting to Communication Styles   248
    Overcoming Communication Problems   251
    Communication Success Strategies   252
HOW DO YOUR PERSONAL RELATIONSHIPS DEFINE YOU? 253
    Relationship Strategies   253
HOW CAN YOU HANDLE CONFLICT AND CRITICISM? 255
    Conflict Strategies   255
    Dealing With Criticism and Feedback   256
WHAT ROLE DO YOU PLAY IN GROUPS? 258
    Being an Effective Participant   258
    Being an Effective Leader   259
APPLICATIONS   *261*

11 MANAGING CAREER AND MONEY 267
Reality Resources

HOW CAN YOU PLAN YOUR CAREER? 268
    Define a Career Path   268
    Map Out Your Strategy   270

Know What Employers Want   271
HOW CAN YOU JUGGLE WORK AND SCHOOL?   272
    Effects of Working While in School   273
    Sources of Job Information   274
WHAT SHOULD YOU KNOW ABOUT FINANCIAL AID?   278
    Student Loans   278
    Grants and Scholarships   280
HOW CAN STRATEGIC PLANNING HELP YOU MANAGE MONEY?   282
    Short-Term Sacrifices Can Create Long-Term Gains   282
    Develop a Financial Philosophy   282
HOW CAN YOU CREATE A BUDGET THAT WORKS?   283
    The Art of Budgeting   284
    A Sample Budget   285
    Saving Strategies   286
    Managing Credit Cards   287
REAL WORLD PERSPECTIVE   289
APPLICATIONS   290

12 MOVING AHEAD   293
Building a Smart Future

WHAT ARE SOME OF THE BIG QUESTIONS IN SCIENCE TODAY?   294
    Something to Prove: Does the Truth Exist?   296
HOW CAN YOU LIVE WITH CHANGE?   296
    Accept the Reality of Change   296
    Maintain Flexibility   297
    Adjust Your Goals   299
WHAT WILL HELP YOU HANDLE SUCCESS AND FAILURE?   300
    Dealing With Failure   300
REAL WORLD PERSPECTIVE   303
    Dealing With Success   304
WHY GIVE BACK TO THE COMMUNITY AND THE WORLD?   305
    Your Imprint on the World   305
    Valuing Your Environment   306
WHY IS COLLEGE JUST THE BEGINNING OF LIFELONG LEARNING?   308
HOW CAN YOU LIVE YOUR MISSION?   309
    Live With Integrity   310
    Roll With the Changes   311
    Learn From Role Models   311
    Aim for Your Personal Best   311
APPLICATIONS   313

Endnotes   321

Index   325

Preface

KEYS TO SCIENCE SUCCESS OWNER'S MANUAL

This book is to provide students having a current or potential science major with realistic and useful steps that will increase their chances for success in college and after college in the workplace. Some hints for *Success* in science follow.

THE ESSENTIALS FOR *SUCCESS*

- Enjoy learning
- See the importance of learning science
- Understand ethical principles and responsibility in science
- Continue learning after graduation
- Use knowledge and skills responsibly
- Graduate from college with a liberal arts education in addition to science
- Complete an internship, perform extra work on projects, or work in the science lab
- Find a mentor

WHAT *SUCCESS* MAY INCLUDE (but not essential conditions)

- High GPA
- A career in science after graduation
- Ability to make a great deal of money
- Highly marketable degree

The Top Three Rules for *Success*

The top three rules for *success* are based on consistent behaviors that lead to gaining a healthy dose of knowledge and skill acquisition while you are in college. The top three rules you will need to follow to succeed as a science major are:

1. Go to class
2. Learn to study
3. Take school seriously (study)

As your authors, we have talked to students across the country. We've learned that you are concerned about your future, you want your education to serve a purpose, you are adjusting to constant life changes, and you want honest and direct guidance on how to achieve your goals. We designed the features of *Keys to Science Success* based on what you have told us about your needs.

The Contents of the Package: What's Included

We chose the topics in this book based on what you need to make the most of your educational experience. You need a strong sense of *self*, *learning style*, and *goals* in order to discover and pursue the best course of study. You need good *study skills* to take in and retain what you learn both in and out of class. You need to *manage your time*, *money*, and *relationships* so you can handle the changes life hands you. *Keys to Science Success* can guide you in all of these areas and more.

The distinguishing characteristics and sections of this book are designed to make your life easier by helping you take in and understand the material you read.

Lifelong learning. The ideas and strategies you learn that will help you succeed in school are the same ones that will bring you success in your career and in your personal life. Therefore, this book focuses on success strategies as they apply to *school*, *work*, and *life*, not just to the classroom or laboratory.

Thinking skills. Being able to remember facts and figures won't do you much good at school or beyond unless you can put that information to work through clear and competent thinking. This book has a chapter on *critical and creative thinking* that will help you explore your mind's actions and thinking processes.

Skill-building exercises. Today's graduates need to be effective thinkers, team players, writers, and strategic planners. The exercises at the end of the chapters will encourage you to develop these valuable career skills and to apply thinking processes to any topic or situation.

Diversity of voice. The world is becoming increasingly diverse in ethnicity, perspective, culture, lifestyle, race, choices, abilities, needs, and more. Every student, instructor, course, and school is unique. One point of view can't possibly apply to everyone. Therefore, many voices will speak to you from these pages. What you read will speak to your needs, offer ideas, and treat you with respect.

User-friendly features. The following features will make your life easier in small but significant ways.

- **Perforations.** Each page of this book is perforated so you can tear out exercises to hand in, take with you somewhere, or keep in your date book as a reference.
- **Exercises.** The exercises are together at the ends of the chapters, so if you want to hand them all in you can do so without also removing any of the text.
- **Definitions.** Selected words are defined in the margins of the text.
- **Long-term usefulness.** Yes, most people sell back some of the textbooks they use. If you take a good look at the material in *Keys to Science Success*, however, you may want to keep this book around. *Keys to Science Success* is a reference that you can return to over and over again as you work toward your goals in school, work, and life.

TAKE ACTION: READ

You are responsible for your education, your growth, your knowledge, and your future. The best we can do is offer some great suggestions, strategies, ideas, and systems that can help. Ultimately, it's up to you to use whatever fits your particular self with all of its particular situations, needs, and wants, and make it your own. You've made a terrific start by choosing to pursue an education—take advantage of all it has to give you.

Acknowledgments

This book has come about through a heroic group effort. We would like to take this opportunity to acknowledge the people who have made it happen. Many thanks to:

- Our student editors Michael Jackson and Aziza Davis.
- Student reviewers Sandi Armitage, Marisa Connell, Jennifer Moe, and Alex Toth.
- Our reviewers: Glenda Belote, Florida International University; John Bennett, Jr., University of Connecticut; Ann Bingham-Newman, California State University, L.A.; Mary Bixby, University of Missouri–Columbia; Barbara Blandford, Education Enhancement Center at Lawrenceville, NJ; Jerry Bouchie, St. Cloud State University; Mona Casady, SW Missouri State University; Janet Cutshall, Sussex County Community College; Valerie DeAngelis, Miami-Dade Community College; Rita Delude, NH Community Technical College; Judy Elsley, Weber State University (Ogden, UT); Gregg R. Godsey, Riverside High School (Washington); Sue Halter, Delgado Community College; Suzy Hampton, University of Montana; Maureen Hurley, University of Missouri–Kansas City; Karen Iversen, Heald Colleges; Kathryn Kelly, St. Cloud State University; Nancy Kosmicke, Mesa State College in Colorado; Frank T. Lyman, Jr., University of Maryland; Barnette Miller Moore, Indian River Community College in Florida; Rebecca Munro, Gonzaga University in Washington; Virginia Phares, DeVry of Atlanta; Brenda Prinzavalli, Beloit College in Wisconsin; Jacqueline Simon, Education Enhancement Center at Lawrenceville, NJ; Carolyn Smith, University of Southern Indiana; Joan Stottlemyer, Carroll College in Montana; Thomas Tyson, SUNY Stony Brook; Rose Wassman, DeAnza College; Michelle G. Wolf, Florida Southern College.
- The PRE 100 instructors at Baltimore City Community College, Liberty Campus, especially college President Dr. Jim Tschechtelin, Coordinator Jim Coleman, Rita Lenkin Hawkins, Sonia Lynch, Jack Taylor, and Peggy Winfield. Thanks also to Prentice Hall representative Alice Barr.
- The instructors at DeVry, especially Susan Chin and Carol Ozee.
- The instructors at Suffolk Community College and Prentice Hall representative Carol Abolafia.
- Our editorial consultant Rich Bucher, professor of sociology at Baltimore City Community College.
- Dr. Frank T. Lyman, inventor of the Thinktrix system.
- Professor Barbara Soloman, developer of the Learning Styles Inventory.
- The people who contributed their stories for Real World Perspectives: Anonymous, Clacy Albert, Laura Brinckerhoff, Brett Cross, Erica Epstein, Norma Espina, Jeff Felardeau, Edith Hall, Jacque Hall, Mike Jackson, Miriam Kapner, Karin Lounsbury, Matt Millard, Todd

Montalbo, Carolyn Christina Moos, Carrie Nelson, Tan Pham, Chelsea Phillips, Patti Reed-Zweiger, Raymond Reyes, Tim Short, Tom Smith, Janis M. Wignall, and Shirley Williamson.

- Kathleen Cole, assistant and student reviewer extraordinaire, and Giuseppe Morella.
- Our editor Sande Johnson.
- Our production team, especially Glenn Johnston, Mary Carnis, Marianne Frasco, Steve Hartner, and Nancy Velthaus.
- The folks in our marketing department, especially Jeff McIlroy, Frank Mortimer, Jr., Karen Austin, Robin Baliszewski, and Christopher Eastman.
- Jackie Fitzgerald, Beth Bollinger, Jennifer Collins, Amy Diehl, Byron Smith, Julie Wheeler, and Robin Diamond.
- The Prentice Hall representatives and the management team led by Gary June.
- Our families and friends.
- Judy Block, who contributed both editing suggestions and study skills text.

Finally, for their ideas, opinions, and stories, we would like to thank all of the students and professors with whom we work. We appreciate that, through reading this book, you give us the opportunity to learn and discover with you.

About the Authors

Janet R. Katz is a cardiac rehabilitation nurse, an adjunct instructor at Gonzaga University, and author of the book *Majoring in Nursing*, as well as various articles on nursing and medicine. She is a contributor to the *Keys to Success in College* modules on scientific research and health science careers. Janet is active in advancing the profession of nursing and its mission of preventing disease, promoting health, and advocating the health care of individuals, families, and communities both locally and globally. After several careers, including that of family planning counselor in central Massachusetts, research assistant at the University of Washington in Seattle, horseshoer in Washington state, and stage technician at Washington State University, Janet became a registered nurse. Janet lives in Spokane, Washington with her husband and their two dogs.

Janet is currently seeking a Ph.D. in education from Gonzaga University. She holds a master's degree in nursing from Gonzaga University, and a bachelor's in nursing from Washington State University, as well as her RNC certification from the American Nurses Association Credentialing body, the ANCC.

Carol Carter is Vice President and Director of Student Programs and Faculty Development at Prentice Hall. She has written *Majoring in the Rest of Your Life: College and Career Secrets for Students* and *Majoring in High School*. She has also co-authored *Graduating Into the Nineties*, *The Career Tool Kit*, *Keys to Career Success*, *Keys to Effective Learning*, and the first edition of *Keys to Success*. In 1992 Carol and other business people co-founded a nonprofit organization called LifeSkills, Inc., to help high school students explore their goals, their career options, and the real world through part-time employment and internships. LifeSkills is now part of the Tucson Unified School District and is featured in seventeen high schools in Tucson, Arizona.

Joyce Bishop holds a Ph.D. in psychology and has taught for more than twenty years, receiving a number of honors, including Teacher of the Year. For the past four years she has been voted "favorite teacher" by the student body and Honor Society at Golden West College, Huntington Beach, CA, where she has taught since 1986 and is a tenured professor. She is currently working with a federal grant to establish Learning Communities and Workplace Learning in her district, and has developed workshops and trained faculty in cooperative learning, active learning, multiple intelligences, workplace relevancy, learning styles, authentic assessment, team building, and the development of learning communities. She also co-authored *Keys to Effective Learning*.

Sarah Lyman Kravits comes from a family of educators and has long cultivated an interest in educational development. She co-authored *The Career Tool Kit*, *Keys to Study Skills*, and the first edition of *Keys to Success* and has served as Program Director for LifeSkills, Inc., a nonprofit organization that aims to further the career and personal development of high school students. In that capacity she helped to formulate both curricular and organizational elements of the program, working closely with instructors as well as members of the business community. She has also given faculty workshops in critical thinking, based on the Thinktrix critical thinking system. Sarah holds a B.A. in English and drama from the University of Virginia, where she was a Jefferson Scholar, and an M.F.A. from Catholic University.

Researching Your Science Education

Collecting the Basic Data

Welcome—or welcome back—to an education in science. Whether you are right out of high school, returning to student life after working for some years, or continuing on a current educational path, you are facing new challenges and changes. Every person has a right to seek the self-improvement, knowledge, and opportunity that an education in science can provide. By choosing to pursue science, you have given yourself a strong vote of confidence and the chance to improve your future.

This book will help you fulfill your potential as a science major by giving you keys—ideas, strategies, and skills—that can lead to success in school, on the job, and in life. Chapter 1 will give you an overview of the science education world. It will start by looking at today's science students—who they are and how they've changed—and at the connection between a science education and success. You will also discover in this chapter how various resources can help you deal with issues and problems and how teamwork plays a role in your success.

In this chapter, you will explore answers to the following questions:

- Why do you need to study a variety of sciences?
- Who is pursuing science education today?
- How do science knowledge and skills promote success?
- What resources are available at your school?
- What is your role in a diverse world?

WHY DO YOU NEED TO STUDY A VARIETY OF SCIENCES?

You probably already know the reasons why a good science background is important: To keep up with rapid advances that affect daily life. With newspaper headlines announcing: "Scientists urge more prudent use of antibiotics" and magazine articles discussing robotics, DNA, and the "geometry lesson of the marching ants," it takes only basic observation to see that life is rapidly changing due to advances in science and technology. If you are 18 years old and just beginning college, think back ten years. What kind of computer did you have? What kind of treatments were available for HIV? If you are older and returning to school, the contrast is much more vivid. Do you remember a time when you didn't own a VCR? Did you have e-mail? Had you even heard of e-mail? Do you remember a time when no one talked about greenhouse gases or global warming? And almost everyone can remember a time when complex genetic engineering and cloning were not occurring.

Examples of how technology and science affects our lives abound, and it is for this reason that a knowledge of science and math skills is needed. You need this knowledge even if you do not pursue a science career; you need it to be an active citizen and a responsible family member. For instance, you must be able to understand the implications, both ethical and practical, of genetic testing and therapy, the spread of viruses, and of disappearing wetlands, rain forests, and other natural habitats. Can you understand the research presented in the articles you read? Can you discern fact from fiction, reality from sensationalism? If you read about a new study on exercise, engines, or equilibrium, can you put it to use?

All of us are called on to make political, social, and personal decisions regarding everything from healthcare to finances, from international foreign aid to environmental protection, and from genetically engineered tomatoes to gene therapy for a host of diseases. The decision to major in science is a good one and one that will be useful to you in many ways. Science and math knowledge and skills teach you critical thinking, creativity, teamwork, and all around good work habits; each one is essential to any kind of career you pursue.

A major purpose of going to college is to broaden your worldview by taking time to study subjects not specifically related to a major, or career goal. The purpose of this book is to help you learn to succeed in the sciences whether you remain in science your entire life or decide in your senior year to become an art major. And by the way, if this is the case, your science background will help you with painting (chemistry), sculpturing (physics and geometry), ceramics (chemistry and physics), and designing jewelry (metallurgy, physics, and anatomy). Remember, your goals may change as you go through college, but what you learn in the physical, life, computer, or engineering sciences, along with math, will help no matter what you decide.

WHO IS PURSUING SCIENCE EDUCATION TODAY?

In various forms, learning took place in the ancient civilizations of Rome, Greece, Byzantium, and Islam. Learning institutions became formalized as universities, similar to those of present day, in medieval Europe as early as the eleventh century. In the early life of the formal university, students and scholars were men, mostly white, seeking religious and intellectual pursuits. Since their inception at that time, universities have evolved, becoming centers for cultural and social inspiration, intellectual growth, and scientific advancements and research.

Because of federal support and a universal understanding that a formal education should be the right of all people regardless of race, creed, color, age, or gender, universities have become extremely diverse, serving over 14 million people per year in the United States alone. The variety of schools includes community colleges offering certificates and associate degrees, technical schools offering certificates and training in specific technical fields, colleges offering baccalaureate degrees, and universities offering both baccalaureate and advanced degrees. In this new technological age, there are even virtual universities where classes are taken on-line.

Today's college students are more diverse than at any time in history. Although many students still enter college directly after high school, the old standard of the student finishing a four-year college education at the age of 22 is a standard no longer. Some students take longer than four years to finish. Some students complete part of their education, pursue other paths for a while, and return to finish later in life. Some go right into the work force after high school and decide to pursue a college degree after many years. The old rules no longer apply in science, technology, or education.

The Diverse Student Body

The following facts from the National Center for Education Statistics paint a dramatic picture of how the student population has changed since the 1980s.

- Twenty-three percent of college students were science majors in 1995. Of these, 6.9 percent majored in the health sciences; 4.8 percent in biological and life sciences; and 1.6 percent in physical sciences and science technologies. The remaining 10 percent studied computer science, engineering science, or mathematics.[1]

- African-American and Asian and Pacific Islander students made up about 14 percent of students majoring in health, biological, and physical sciences in 1994–95; Hispanic, American Indian, and Alaskan Native students 0.5 percent.[2]
- Significantly more women than men students earned bachelor's degrees in health sciences. On the other hand, substantially more male than female students earned degrees in computer science, engineering science, or technology.[3]
- Most science majors younger than age 25 majored in biological sciences. The majority of science students between the ages of 25 and 34 majored in medicine, dentistry, or other health-related fields, and those older than 35 years old focused their majors on the health sciences.[4]
- Of the entire undergraduate population in the academic year 1995–96, 20 percent were supporting children. Among older students (entering as a freshman at the age of 20 or older), more than 40 percent were supporting children.[5]
- Almost 80 percent of undergraduates were employed at some time during their enrollment in the academic year 1995–96. Of students who worked, 26 percent worked full-time; 19 percent also attended school full-time.[6]
- Students are taking longer to get a degree. Of students graduating in 1995, 64.5 percent took more than four years to complete their degrees, and 25.6 percent took more than six years.[7]

"A journey of a thousand miles must begin with a single step."
LAO-TZU

These changes have brought with them a new education. experience defined by the varying needs of an increasingly diverse student body. Not so long ago, if you were female, African-American, or disabled, you had limited opportunity to attend college, much less major in science. Even twenty years ago, you might have given up on an education if you couldn't afford a four-year college, if you were unable to attend classes during the day, or hadn't enrolled immediately after high school and then felt it was too late to return. Now, however, a science education is available to a wide range of potential students, regardless of their situations or backgrounds.

A science education isn't an automatic guarantee of a high-level, high-paying job. Statistically, however, a better-educated population means a more efficient work force, more career fulfillment, and better-paid workers. Quality of life can improve when people make the most of their abilities through education.

The decision to take advantage of an education in science is in your hands. You are responsible for seeking out opportunities and weaving school into the fabric of your life. You may face some of these challenges:

- Handling the responsibilities and stress of parenting children alone, without a spouse
- Returning to school as an older student and feeling out of place
- Learning to adjust to the cultural and communication differences in the diverse student population

- Having a physical disability that presents challenges
- Having a learning disability such as dyslexia or attention-deficit hyperactivity disorder (ADHD)
- Balancing a school schedule with part-time or even full-time work
- Handling the enormous financial commitment college requires

Your school can help you work through these and other problems if you actively seek out solutions and help from available support systems around you. Explore some reasons why the hard work is worthwhile.

Women in Science

The U. S. Bureau of Labor Statistics estimates that overall entry into the labor market will decline between 1994 and 2005. But women entering the work force will grow twice as fast as men.[8] This means that by the year 2005 about two-thirds of all women will be working; however, female science majors do not enter the fields of engineering, math, or computer science as frequently as men do.[9]

This leads the National Science Foundation to express concern about the following:[10]

- A disportionately large number of girls lose interest in science during elementary and middle school.
- Low numbers of high school women enroll in advanced science and math courses to prepare for college.
- Low numbers of women are entering or completing undergraduate programs in science, engineering, and math.
- There is a slow rate of advancement of women to senior-level leadership positions in academia, industry, business, and government careers.

There are many resources for women considering science, math, and technology careers. Many professional organizations have branches that are exclusively for women; there are many books, articles, and on-line resources as well. Grants and scholarship money are available through professional organizations and through the National Science Foundation, which has the promotion of girls' and women's education at the top of their agenda.

HOW DO SCIENCE KNOWLEDGE AND SKILLS PROMOTE SUCCESS?

There are many reasons for making a strong, if not outright desperate, case for intensive science education efforts. Four of these reasons include:

1. U. S. Department of Education Secretary Richard W. Riley, along with national economic experts, and the U. S. Bureau of Labor Statistics (BLS) emphasize that advanced knowledge and skills in science, math,

and technology are critical to the long-term economic well-being of the United States.[11]
2. Secretary Riley further states that thousands of job applicants are turned down due to a lack of math, science, communication, and computer proficiency.[12]
3. The BLS reports that the two fastest-growing job areas, computer science and technology and health sciences, require mastery of both science and math.[13]
4. Employment data shows that those with higher-level math and science skills make significantly more money and have fewer and shorter periods of unemployment than those with similar levels of other types of education.[14]

There is, of course, no guarantee that a job will await you once you leave college with a science degree. The statistics cited by the BLS and others are predictions, not facts. For some local insight take a look at your area newspaper's classified want ads. Where are the jobs? Are there six pages of clerical and unskilled labor jobs and one page of professional and technical jobs? Go to the library and look at newspapers from across the country, and in the reference section, refer to the *Occupational Outlook Handbook* and *Peterson's Guide to Jobs*. Ask the reference librarian to help you find information on science careers to get a clearer picture of the job market. Another tactic is to look at the many Web sites that discuss science careers, such as Kaplan's www.Kaplan.com, where you can find information not only on careers, but also colleges and college testing.

If, however, you find that the career you want is rated poorly for job opportunities, don't give it up. There will always be job openings in every field as people retire, move, change careers, or get fired. If you want to be a

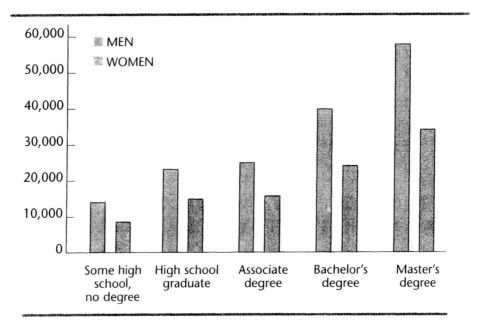

Figure 1-1

Education and income.

Source: U. S. Department of Commerce, Bureau of the Census, *Current Population Reports*, Series P-60, *Monthly Income of Households, Families, and Persons in the United States: 1994*.

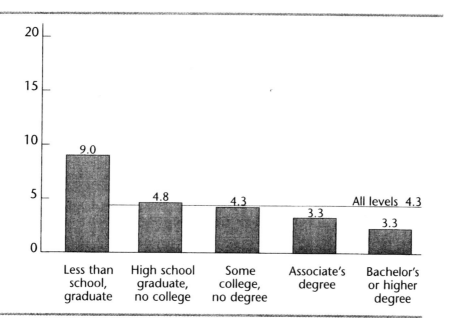

Figure 1-2

Education and employment.

Source: U. S. Department of Commerce, Bureau of the Census, *Current Population Reports*, Series P-60, *Monthly Income of Households, Families, and Persons in the United States: 1994.*

marine biologist working with dolphins don't give it up just because the paying jobs are few and far between. If you really want to work in a specific and highly specialized field, study hard and understand that it may take you longer, for instance, than your engineering science classmate to find a job. New positions are also opening up in all fields as new discoveries are made. NASA administrator Daniel S. Goldin, speaking to the Commonwealth Club of California, talked about the use of virtual reality to help test products and machines before they are built, train surgeons in new surgical techniques, understand space travel, and in conjunction with biological principles, understand the superior efficiency of how brain cells hold vast amounts of information and apply it to computers. The cutting edge science and technology you learn today will benefit you tomorrow.

Education improves your quality of life and expands your self-concept. Income and employment get a boost from education. *The Digest of Education Statistics 1996* reports that income levels rise as educational levels rise. Figure 1-1 shows average income levels for different levels of educational attainment. Figure 1-2, also from a report in the *Digest*, shows how unemployment rates decrease as educational levels rise.

Understanding is joyous.
CARL SAGAN

As you rise to the challenges of education, you will discover that your capacity for knowledge and personal growth is greater than you imagined. As your abilities grow, so do opportunities to learn and do more in class, on the job, and in your community.

All education increases your possibilities. Education gives you a *base of choices* and *increased power*, as shown in Figure 1-3. First, through different courses of study, it introduces you to *more choices* of career and life goals. Second, through the different types of training you receive, it gives you *more power* to achieve the goals you choose. For example, while taking a writing

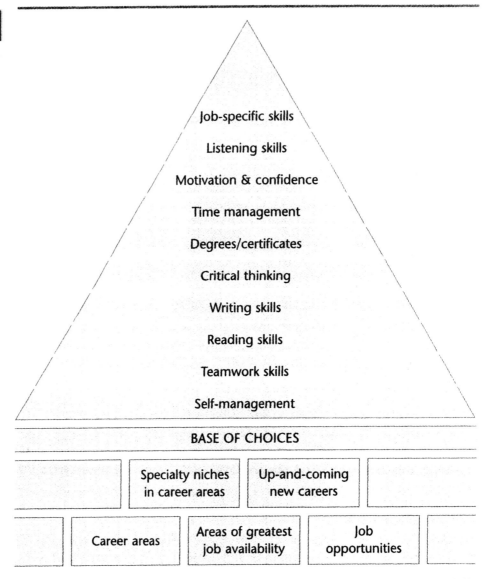

Figure 1-3
Education increases choices and power.

class, you may learn about careers in journalism. This experience may lead you to take a class in journalistic writing that teaches you about science reporting. Down the road, you may decide to work on a newspaper as a science writer and to make science journalism your career. Looking back, you realize that two classes you took in college changed the course of your life.

A good science education improves your employability and earning potential. Learning additional skills raises your competency so you can fulfill the requirements of higher-level jobs. In addition, having a college degree in science makes an impression on potential employers.

Science education also makes you a well-rounded person as it widens your understanding about what is possible in the world. Science increases your awareness and appreciation of areas that affect and enrich human lives, such as music, art, literature, politics, and economics.

Your science education affects both community involvement and personal health. Education helps to prepare individuals for community activism by helping them understand political, economic, and social conditions. Education also increases knowledge about health behaviors and preventive care. The more education you have, the more likely you are to practice healthy habits in your daily life and to make informed decisions.

A science education is more than the process of going to school and earning a degree or certificate. It is a choice to improve your mind and your skills. Any program in science, no matter the length or the focus, is an opportunity to set and strive for goals. If you make the most of your mind, your time, and your educational opportunities, you will realize your potential. Using available resources is part of that process.

WHAT RESOURCES ARE AVAILABLE AT YOUR SCHOOL?

Resources help you make the most of your education. As a student, you are investing money and time. Whether you complete your studies over the course of six months or sixty years, resources can help you get where you want to go.

Like any company that makes products or provides services, your school is a business. The goal of this particular business is the successful education of all who come through its doors. Table 1-1 offers a general summary of resources, most or all of which can be found at your school. Most schools offer a student orientation, near the beginning of your first semester, that will explain resources and other important information. Even if your school does not, you can orient yourself. The following sections will describe helpful resources—people, student services, organizations, and course catalogs and student handbooks.

Resources People, organizations, or services that supply help and support for different aspects of college life.

People

Your school has an array of people who can help you make the most of your educational experience: instructors, administrative personnel, advisors and counselors, and teaching assistants. They're often busy with numerous responsibilities, but their assistance is provided as a standard part of your educational package. Take the opportunity to get to know them, and to let them get to know you. Together you can explore how they can help you achieve your goals.

Instructors are more than just sources of information during scheduled class time. Instructors can also become your mentors. Mentors are people who will help guide you as you develop yourself as a scientist. Finding a mentor involves the following:

- Get to know your instructors by working hard in labs and classes and by assisting in research projects. You can do this through work-study programs, teaching assistantships, or by volunteering. Read Bob Weinstein's book, *I'll Work for Free*, for other tips on finding such opportunities.

- Get hands-on experience in science. This is crucial to understanding what the field is all about. It may include counting the tiny bones of

Table 1-1: How resources can help you.

RESOURCE	ACADEMIC ASSISTANCE	FINANCIAL ASSISTANCE	JOB/CAREER ASSISTANCE	PERSONAL ASSISTANCE
Instructors	Choosing classes, clarifying course material, help on assignments, dealing with study issues		Can tell you about their fields, may be a source of networking contacts	During office hours, are available to talk to you
Administrators	Academic problems, educational focus, problems with school services		Can be a source of valuable contacts	Can help you sort through personal problems with instructors or other school employees
Academic Advisors	Choosing, changing, or dropping courses; getting over academic hurdles; selecting/changing a major		Can advise you on what job opportunities may go along with your major or academic focus	
Personal Counselors	Can help when personal problems get in the way of academic success	Services may be free or offered on a "sliding scale" depending on what you can afford		Help with all kinds of personal problems
Financial Aid Office		Information and counseling on loans, grants, scholarships, financial planning, work-study programs	Information on job opportunities within your school environment (work-study program and others)	
Academic Centers	Help with what the center specializes in (reading, writing, math)		Perhaps an opportunity to work at the center	
Organizations and Clubs	If an academic club, can broaden your knowledge or experience in an area of study; can help you balance school with other enriching activities		Can help you develop skills, build knowledge, and make new contacts that may serve you in your working life	Depending on the club focus, can be an outlet for stress, a source of personal inspiration, a source of important friendships, an opportunity to help others
Fitness Center(s)		Usually free or low cost to enrolled students		Provides opportunity to build fitness and reduce stress; may have weight room, track, aerobic or dance classes, martial arts, team sports, exercise machines, etc.

Table 1-1 Continued.

RESOURCE	ACADEMIC ASSISTANCE	FINANCIAL ASSISTANCE	JOB/CAREER ASSISTANCE	PERSONAL ASSISTANCE
Bulletin Boards	List academic events, class information, changes and additions to schedules, office hours, academic club meetings	List financial aid seminars, job opportunities, scholarship opportunities	List career forums, job sign-ups, and employment opportunities; offer a place for you to post a message if you are marketing a service	List support group meetings
Housing and Transportation Office		Can help find the most financially beneficial travel or housing plan		Can help commuters with parking, bus or train service, and permits; can help with finding on- or off-campus housing
Career Planning and Placement Office		Can help add to your income through job opportunities	Job listings, help with résumés and interviews, possible interview appointments, factual information about the workplace (job trends, salaries, etc.)	
Tutors	One-on-one help with academic subjects; assistance with specific assignments		If you decide to become a tutor, a chance to find out if teaching and working with people is for you	
Student Health Office		May provide low-cost or no-cost health care to enrolled students; may offer reduced-cost prescription plan		Wellness care (regular examinations), illness care, hospital and specialist referrals, and prescriptions
Adult Education Center	Academic help tailored to the returning adult student	May have specific financial-aid advice	May have job listings or other help with coordinating work and classes	May offer child-care assistance and opportunities to get to know other returning adults
Support Groups and Hotlines	If school-related, they offer a chance to hear how others have both stumbled and succeeded in school— and a chance to share your story			Personal help with whatever the hotline or support group specializes in; a chance to talk to someone whose job is to listen
School Publications	Academic news and course changes	News about financial aid opportunities or work-study programs	Job listings, information about the workplace and the job market	Articles and announcements about topics that may help you

bats, doing library reference work, observing and tracking numbers of migrating birds, entering data on a computer, washing laboratory glassware, preparing petri dishes, or setting up for lab classes. Whatever it takes, do this kind of work to get to know your instructors and the work they do.

◆ Work hard to show that you have an interest in the instructor's field. You can help clarify your interest in that field and at the same time develop a relationship with an expert.

The value of a mentor extends beyond finding someone to help guide you while you are in school. A mentor can help you with job and graduate school recommendations through connections they have with other scientists around the country.

In this book the term "instructor" is used for simplicity's sake, but instructors have official titles that show their rank. From lowest to highest, these include lecturer, instructor, assistant professor, associate professor, and full professor (often just called professor).

Instructors have many time-draining responsibilities outside of teaching. However, you can gain access to your instructors while still respecting the demands on their time. Most instructors keep office hours, and they will tell you the location and times. You are responsible for seeking out your instructor during office hours. If your schedule makes this impossible, let your instructor know. Perhaps you and your instructor can get together at another time. Most schools have electronic mail (e-mail) systems that allow you to send messages via computer; you may be able to communicate with your instructor using e-mail.

Teaching assistants are people who help an instructor with a course. You may or may not have teaching assistants in your courses. Often they are studying to be instructors themselves. Sometimes they teach the smaller discussion sections that accompany a large group lecture. They can be a great resource when your instructor is too swamped to talk to you.

Your school's *administrative personnel* have the responsibility of delivering to you—the student consumer—a first-rate product. That product is the sum total of your education, comprising facilities, instructors, materials, and courses. Schedule a meeting with your dean, the chair of a particular department, or other school administrator if you have an issue to discuss, such as a conflict with an instructor, an inability to get into a class you need, or a school regulation that causes a problem for you. Although administrators don't interact with students as often as instructors do, it is their business to know how the school is serving you.

Advisors and counselors can help with both the educational and personal sides of being a student. They provide information, advice, a listening ear, referrals, and other sources of help. Generally, students are assigned academic advisors with whom they meet at least once a semester. Your academic advisor will help you find out about classes, choose a schedule, explore and select a major or design your own when the time comes, and plan out the big picture of your academic life. Visit your academic advisor more than once a semester if you have questions or want to make changes.

Counselors, although not usually assigned, are available to you at any time, usually through student services. Don't hesitate to seek a counselor's help if you have something on your mind. The ups and downs of your personal life greatly influence the quality of your work in school. If you put some effort into working through personal problems, you will be more able to do your work well and hand it in on time. Occasionally, an illness or family difficulty may interfere with your schoolwork enough to call for special provisions for the completion of your classes. Most colleges are more than happy to assist you during challenging times.

Student Services

Your school has a variety of services aimed at helping students. Basic services offered by almost every school include the following: academic advising and personal counseling, student health/wellness, career planning and placement, tutoring, fitness/physical education, and financial aid. Depending on your school, you may also find other services: child care, housing and transportation, adult education services (for adults returning to school), disabled student services, academic centers (reading center, writing center, math center, etc.—for help with these specific subjects), various support groups, and school publications that help keep you informed of developments that affect you.

Often a school will have special services for specific populations. For example, at a school where most of the students commute, there may be a transportation office that helps students locate bus schedules and routes, find parking and sign up for permits, or track down car pools. Similarly, at a school where many students are parents, a child-care center may provide day care during class time and also refer students to outside babysitters. You will find additional details about school services in Table 1-1 on pp. 10–11. They can help you earn the maximum benefit from your educational experience.

Organizations

No matter what your needs or interests are, your school probably has an organization that would interest you or can help you. Some organizations are sponsored by the school (academic clubs), some are independent but have a branch at the school (government ROTC programs), and some are student-run organizations (Latino Student Association). Some organizations focus on courses of study (Nursing Club), some are primarily social (fraternities and sororities), some are artistic (Chamber Orchestra), and some are geared toward a hobby or activity (Runner's Club). Some you join in order to help others (Big Brothers or Big Sisters), and some offer help to you (Overeaters Anonymous).

When you consider adding a new activity to your life, weigh the positive effects against the negative effects. Positive effects could be new friends, fun activities, help, a break from schoolwork, stress relief, improved academic performance, increased teamwork and leadership skills, aid to others, and experience that can broaden your horizons. On the negative side, there may be a heavy time commitment, dues, inconvenient location or meeting times, or too much responsibility. Explore any club carefully to see if it makes sense for you. As you make your decision, consider this: Studies have shown that students who join organizations tend to persist in their educational goals more than those who don't branch out.

To find out about organizations at your school, consult your student handbook, ask friends and instructors, or check the activities office or center if your school has one. Some schools, on registration days, have an area where organizations set up tables and make themselves available to talk to interested students. Some organizations seek you out based on your academic achievements. Find out as much as you can. Ask what is expected in terms of time, responsibility, and any necessary financial commitment. Talk to students who are currently involved. Perhaps give an organization a test run to see if you like it.

If you try out an organization, make a commitment that you will stay for the right reasons. Don't be afraid of being labeled a "dropout"; if something becomes more than you can handle, bow out gracefully. In the best of all possible worlds, your involvement in organizations will enrich your life, expand your network of acquaintances, boost your time management skills, and help you achieve goals.

Science Student Organizations

Depending on the size of your school, there will be many different types of science student organizations. Larger schools will have a great deal of variety because they offer more varied science courses. However, all schools will have at least one science club or association. For instance, Purdue University has an Association of Minority Science Students, a Math Club, a Geoscience Society, a Science Student Council, and the University Meteorological Association. Washington State University has an Astronomy Club and a Physics Club, and Harvard University has a Computer Society, a Health Awareness Advocacy Group, Women in Science, and even a Science Fiction Association.

Check your school's Web site, under students and student organizations. If your interests are not represented, you can always start your own group by following the school's protocol for starting new student organizations.

College Catalogs and Student Handbooks

Navigating through your school's course offerings, the departments and resource offices, and even the layout of the campus can seem overwhelming. There are two publications that can help you find your way—the college catalog and the student handbook. Most schools provide these materials as a standard part of their enrollment information.

The *college catalog* is your school's academic directory. It lists every department and course available at the school. Each course name will generally have two parts. Take "EN101" or "CHEM205," for example. The first part is one or more letters indicating department and subject matter, and the second part is a number indicating course level (lower numbers for introductory courses and higher numbers for more advanced ones). The catalog groups courses according to subject matter and lists them from the lowest-level courses up to the most advanced, indicating the number of credits earned for each class. See Figure 1-4 for a segment of an actual college catalog from Washington State University[15]. A course book released prior to each semester will indicate details such as the instructor, the days the course meets, the time of day, the location (building and room), and the maximum number of students who can take the course.

Figure 1-4 Sample page from a course catalog.

MATHEMATICS & COMPUTER SCIENCE – 77

MATH 204 Mathematics for Elementary Teachers II   3 credits
Does not satisfy the university core requirement in mathematics unless the student earns a Teacher Certificate in Elementary Education. This course is the continuation of MATH 203. Topics include introductory geometry, geometric constructions, congruence, similarity, measurement, coordinate geometry, probability and statistics. Prerequisite: MATH 203. *Fall and spring.*

MATH 231 Discrete Structures   3 credits
Topics taken from sets, functions, matrices, ordered sets, partially ordered sets, directed graphs, algebraic systems, recursive definitions, and algorithms. Prerequisite: MATH 258. *Fall.*

MATH 258 Calculus and Analytic Geometry II   4 credits
Introduction to exponential, logarithmic, and hyperbolic functions; parametric equations; polar coordinates; techniques and applications of integration. Prerequisite: MATH 157 (with a grade of C- or higher strongly recommended). *Fall and spring.*

MATH 259 Calculus and Analytic Geometry III   4 credits
Infinite series, vectors, partial derivatives, multiple integrals, solid analytic geometry including spherical and cylindrical coordinates. Prerequisite: MATH 258 (with a grade of C- or higher strongly recommended). *Fall and spring.*

MATH 260 Ordinary Differential Equations   3 credits
Solution methods for first order equations, and for second and higher order linear equations. Includes series methods, and solution of linear systems of differential equations. Prerequisite: MATH 259 (with a grade of C- or higher strongly recommended). *Fall and spring.*

MATH 281 Introduction to Abstract Mathematics   3 credits
Introduction to standard proof techniques used in abstract mathematics. The concepts of implications, converses, contrapositives, direct proofs, indirect proofs, proof by contradiction, and existence and uniqueness theorems will be considered. Topics will be chosen from logic, set theory, relations and functions, cardinality, number theory, algebra, and analysis. Prerequisite: MATH 259 or permission. *Spring.*

MATH 290 Sophomore Directed Reading   credit by arrangement
Readings and reports in selected mathematical topics. Prerequisite: permission. *On sufficient demand.*

Upper Division

MATH 321 Statistics for Experimentalists   3 credits
An extensive survey of statistical methods with an emphasis on their application. The focus is on inferential statistics with topics chosen from estimation, significance tests, regression, correlation, analysis of variance, multiple regression, nonparametric statistics, goodness of fit tests, and contingency tables. Prerequisite: MATH 157 or permission. *Fall and spring.*

MATH 328 Operations Research   3 credits
Quantitative methods for application to problems from business, engineering, and the social sciences. Topics include linear and dynamic programming, transportation problems, network analysis, PERT, and game theory. Prerequisite: MATH 258. *Spring, even years.*

MATH 339 Linear Algebra   3 credits
A systematic study of matrices, vector spaces, and linear transformations. Topics include systems of linear equations, determinants, dependence, bases, dimension, rank, eigenvalues and eigenvectors. Applications include geometry, calculus, and differential equations. Prerequisite: MATH 259. *Spring.*

MATH 341 Modern Geometry   3 credits
Axiomatic systems for, and selected topics from, Euclidean geometry, projective geometry, and other non-Euclidean geometries. Special attention will be given to the needs of the individuals preparing to teach at the secondary level. Prerequisite: MATH 259. *Spring.*

MATH 350 Elementary Numerical Analysis   3 credits
An introduction to numerical analysis: root finding, interpolation, numerical integration and differentiation, finite differences, numerical solution to initial value problems, and applications on a digital computer. Prerequisite: MATH 258. *Spring, odd years.*

MATH 371 (ENSC 371) Advanced Engineering Mathematics   3 credits
Application of advanced mathematical techniques to problems of interest to engineers, scientists, and applied mathematicians. Analytical methods involving linear algebra, complex variables, and partial differential equations, including the study of special functions. Prerequisite: MATH 260. *Fall and spring.*

MATH 413 Advanced Calculus I   3 credits
Notions from set theory, the real number system, topology of the real line, continuity (including uniform continuity), differentiation, Riemann integration, sequences, and infinite series of numbers and functions. Prerequisite: MATH 281, or MATH 259 and MATH 231, or permission. *Fall, even years.*

MATH 414 Advanced Calculus II   3 credits
Continuation of MATH 413. Prerequisite: MATH 413. *Spring, odd years.*

MATH 417 Complex Variables   3 credits
Complex numbers and functions, analyticity and the Cauchy-Riemann equations, integration, and Cauchy's theorem and formula. Other topics chosen from Taylor and Laurent series, the calculus of residues, conformal mapping, and applications. Prerequisite: MATH 281, or MATH 259 and MATH 231, or permission. *Fall, odd years.*

MATH 421 Probability Theory   3 credits
A mathematical treatment of the laws of probability with emphasis on those properties fundamental to mathematical statistics. General probability spaces, combinatorial analysis, random variables, conditional probability, moment generating functions, Bayes law, distribution theory, and law of large numbers. Prerequisite: MATH 281, or MATH 259 and MATH 231, or permission. *Fall, even years.*

MATH 422 Mathematical Statistics   3 credits
An examination of the mathematical principles underlying the basic statistical inference techniques of estimation, hypothesis testing, regression and correlation, nonparametric statistics, analysis of variance. Prerequisite: MATH 421. *Spring, odd years.*

MATH 437 Abstract Algebra I   3 credits
A detailed examination of topics chosen from groups, rings, integral domains, Euclidean domains, unique factorization, fields, Galois theory, and solvability by radicals. Prerequisite: MATH 281, or MATH 259 and MATH 231, or permission. *Fall, odd years.*

MATH 438 Abstract Algebra II   3 credits
Continuation of MATH 437. Prerequisite: MATH 437. *Spring, even years.*

MATH 450-453 Selected Topics   1-3 credits
Possible topics include combinatorics, topology, number theory, advanced numerical analysis, advanced linear algebra, theory of computation and complexity, and history of mathematics. Credit by arrangement. Prerequisite: third year standing and permission. *On sufficient demand.*

Source: 1997–1999 Undergraduate Catalog, Gonzaga University, Spokane, WA.

Your college catalog contains a wealth of other information. It may give general school policies such as admissions requirements, the registration process, and withdrawal procedures. It may list the departments to show the range of subjects you may study. It may outline instructional programs, detailing core requirements as well as requirements for various majors, degrees, and certificates. It may also list administrative personnel as well as faculty and staff for each department. The college catalog is an important resource in planning your academic career. When you have a question, consult the catalog first before you spend time and energy looking elsewhere.

Your *student handbook* looks beyond specific courses to the big picture, helping you to navigate student life. In it you will find some or all of the following, and maybe more: information on available housing (for on-campus residents) and on parking and driving (for commuters); overviews of the support offices for students, such as academic advising, counseling, career planning and placement, student health, disabled student services, child care, financial aid, and individual centers for academic subject areas such as writing or math; descriptions of special-interest clubs; and details about library and computer services. It may also list hours, locations, phone numbers, and addresses for all offices, clubs, and organizations.

Your student handbook will also describe important policies such as how to add or drop a class, what the grading system means, campus rules, drug and alcohol policies, what kinds of records your school keeps, safety tips, and more. Keep your student handbook where you can find it easily, in your study area at home or someplace safe at school. The information it gives you can save you a lot of trouble when you need to find out information about a resource or service. If you call for locations and hours before you visit a particular office, you'll avoid the frustration of dropping by when the office is closed.

Making the most of your resources is one way to adjust to your new environment. Interacting with people around you is another.

WHAT IS YOUR ROLE IN A DIVERSE WORLD?

It is no news that we are moving toward a global economy, nor that life is becoming increasingly complex. The turn of the century is heralding predictions of vast change as we move from the industrial age of the past 200 years to the knowledge, or information, age. Science and technology have significantly contributed to these changes with advances in computers, communications, and transportation systems. What in the past affected only small regional groups of people, now affects us all. For example, war and civil unrest in the Sudan affects politics throughout world. Japan's or Russia's economic problems cause fluctuations in the U. S. stock market.

Diversity Is Real in Science

What this means for you as a science student is that you will be a part of a global structure that supports, questions, and regulates scientific research policy. As a scientist of any kind, you will not be working in a vacuum; you will have contact with a variety of people even if you work in a small laboratory miles from anyone or anything. You will still need to write grants, present your findings, and communicate to get supplies. Furthermore, the people you contact will not always be from the same culture as you. This is not a choice anymore, but a reality. Part of your science education is learning about diversity and, more importantly, participating in it. You can accomplish this by

- Meeting and working with other students in your classes that are from different ethnic or cultural backgrounds than you are.
- Taking courses on multiculturalism.
- Traveling to other countries to study or to visit.
- Reading books, fiction and nonfiction, that describe the perspectives of people that have grown up in different circumstances than you.
- Watching foreign movies or those made by minority groups (e.g., *Smoke Signals*, 1998, is an award-wining movie produced, directed, written, and acted by American Indians.)
- Keeping up with international news
- Learning a foreign language.

Ethnocentrism

When groups of people believe that their way of thinking is the only way, or a better way than anyone else's, they are being ethnocentric. Ethnocentrism creates an opinion that one's particular group is better than anyone else's. It's important to be proud of your identity, but it's one thing to think your group is terrific and another thing to think that your group is superior to all other groups.

A group can be organized around any sort of uniqueness—the same skin color, accent, country of origin, ideas, interests, religion, traditions, and much more. The problem arises when celebrating your own uniqueness leads to putting down someone else's. One example is thinking that when someone speaks with an accent, he or she doesn't know as much as you do. Another example is thinking that it is disrespectful for someone not to look you in the eye during a conversation. In certain other cultures, it is considered rude to look people in the eye, especially if that person happens to be an authority.

Ethnocentrism has many negative effects. It can get in the way of effective communication, as you will see in more detail when you read Chapter 10. It can prevent you from getting to know people from different backgrounds. It can result in people being shut out and denied opportunities that all people deserve. It limits you and your potential because it denies you exposure to new ideas that could help you grow and learn. Finally, it can hinder your ability to work with others, which can cause problems for you both at school and on the job.

> **TERMS**
> **Ethnocentrism**
> The condition of thinking that one's particular ethnic group is superior to others.

Diversity and Teamwork

Much of what people accomplish they owe to teamwork. Think of the path of your accomplishments, and you will find that other people had roles in your success. When you earn a degree, complete a project, or raise a family, you don't do it alone. You are part of many hard-working teams. As the African proverb goes, it takes an entire village to raise a child.

REAL WORLD PERSPECTIVE

Brian Roy, Colorado School of Mines, Physics Major

My motivation to become a scientist came through experiencing the monotony of doing remedial labor in the military for two years and in the navy for six years. I wanted to have a job where I wouldn't be doing the same tasks every day and where I could discover new things, ideas, and concepts. One person who helped guide and motivate me along the way was a professor here who encouraged me to study physics for my undergraduate degree and then enter a one-year master's program in materials science. In general, I like how all the professors here have an open-door policy, so you can come and talk to them at any time. Although I like physics, going to one year of graduate school in materials science provides more job opportunities for me. Physics is pretty limited with a master's degree. What really helped me to decide on my course of study and industry direction was talking to a lot of people. I talked to students in different departments and to professors who would give me their five-minute take on life.

Advice

I think that being a scientist, a pure scientist, requires an open mind in every sense, so you are always questioning things. You should also try doing what I'm doing now, that is, working in a physics lab analyzing samples and getting exposure to research scientists like I'm experiencing here at a university. Also, I think people studying science should value working in groups. I'm against it usually, but lately I've been appreciating teamwork. You learn a lot more, and have to explain and talk to someone else.

Your success at school and at work depends on your ability to cooperate in a team setting. At school you will work with study groups, complete group projects, interact with instructors and administrators, and perhaps live with a roommate. At work you will regularly team up with co-workers to achieve goals. At home you work with family or housemates to manage the tasks and responsibilities of daily life. Your achievements depend on how you communicate, share tasks, and develop a common vision.

Any team will gain strength from the diversity of its members. In fact, diversity is an asset in a team. Consider a five-person basketball team, composed of a center, a power forward, a small forward, a shooting guard, and a point guard. Each person has a different role and a different style of play, but only by combining their abilities can they achieve success. The more diverse the team members, the greater the chance that new ideas and solutions will find their way to the table, increasing the chances of solving any problem. As a member of any team, use these three strategies to maximize team success.

1. Open your mind and accept that different team members have valuable roles.
2. Consider the new information and ideas that others offer.
3. Evaluate contributions based on how they help solve the problem or achieve the goal instead of based on the identity of the person who had the idea. Successful teams use what works.

Living Your Role

It's not always easy to open your mind to differences. However, doing so can benefit both you and others around you. You may consider actions like these as you define your role in the diverse world:

- **To accept diversity as a fact of life.** The world will only continue to diversify. The more you adapt to and appreciate this diversity, the more enriched your life will be. Diversity is an asset, not a deficiency.
- **To explore differences.** Open your mind and learn about what is unfamiliar around you.
- **To celebrate your own uniqueness as well as that of others.** It's natural to think that your own way is the best way. Expand your horizons by considering your way as one good way and seeking out different and useful ways to which other people can introduce you.
- **To consider new perspectives.** The wide variety of ideas and perspectives brought by people from all different groups and situations creates a wealth of thought from which the world can find solutions to tough and complex problems.
- **To continue to learn.** Education is one of the most productive ways to combat discrimination and become more open-minded about differences. Classes such as sociology and ethics can increase your awareness of the lives, choices, and values of people in other cultures. Even though your personal beliefs may be challenged in the process, facing how you feel about others is a positive step toward harmony among people.

Throughout this book you will find references to a diverse mixture of people in different life circumstances. Chapter 10 will go into more detail about communicating across lines of difference and addressing the problems that arise when people have trouble accepting each other's differences. Diversity is not a subject that you study at one point in the semester and then leave behind. It is a theme that touches every chapter in this book and every part of your life. Note especially the "Real World Perspective" feature in some chapters, which often highlights people from different backgrounds who are striving to learn about themselves and their world.

"He has not yet learned the lesson of life who does not every day surmount a fear."

RALPH WALDO EMERSON

In Chinese writing, this character has two meanings: One is "chaos"; the other is "opportunity." The character communicates the belief that every challenging, chaotic, demanding situation in life also presents an opportunity. By responding to challenges in a positive and active way, you can discover the opportunity that lies within the chaos.

Let this concept reassure you as you begin college. You may feel that you are going through a time of chaos and change. Remember that no matter how difficult the obstacles, you have the ability to persevere. You can create opportunities for yourself to learn, grow, and improve.

Chapter 1 Applications

Name _____ Date _____

KEY INTO YOUR LIFE
Opportunities to Apply What You Learn

Internet Exercise

You will be using the Internet as you progress in school, and if you don't feel comfortable using it now, you will soon! This exercise is intended to introduce you to a few of the many Internet sites, including one on how to evaluate Internet resources.

Step 1. Get together with a group of two to four other students and go to the library or computer lab.

Step 2. Once you get on the Internet go to the following site:

http://www.mitretek.org/hiti/showcase/documents/criteria.html

Scroll down to "2. Quality Criteria for Evaluation of Health Information" and write down the information under criteria "C1 Credibility." Use this to evaluate a science Internet source.

Step 3. Use the link and go to "Agency for Health Care Policy Research."

Step 4. From here go to "Research Findings," and pick a topic.

Step 5. Using the criteria listed from the evaluation site (the criteria is for health information but may be used in other cases), apply it to the information you found. Is it credible? How do you know?

Step 6. Are there links to any other science Web sites of interest to you?

Skills Analysis, or "I'll never forget the time..."

One method for discovering what skills and important interests you have is by telling a story from your life. Start thinking of a time you did something that was fascinating, significant, or in any way particularly memorable. It doesn't have to be anything that seems connected to science. Begin with the statement: "I'll never forget the time I..." and fill in the rest. You can tell another person your story or write it down. Ask yourself the following questions:

◆ What was so important to you about this event?

- What underlying feelings and thoughts were associated with it?
- What skills, such as observation, reaction, communication, caution, or humor, did you use?

Write down the answers to these questions, and explore ways they might be connected to a scientific field of study.

Career Analysis, or "My three top careers would be..."

This is another complete-the-sentence exercise to help loosen up your brain and get your thoughts going (often referred to as brainstorming).

1. List one career that interests you more than any other. Next, list two more.

2. List the name of one person you can think of in each of the three careers. If you can't think of anyone, write down someone you think could help find such a person, for instance, a mentor, a parent, a reference librarian, or a teacher.

3. Find out how to contact these people either using a phone book or an Internet search. Many people in science careers work for universities, so you can find a way to contact them through the school's Web site, or if they are well known, use a search engine such as Hotbots or Yahoo.

4. Call or e-mail each person, and set up an appointment to meet with them in person, or on the Internet if that is more convenient for them.

5. Ask them the following questions and add some of your own:
    - What is the most interesting part of your work?
    - The least interesting?
    - If I wanted to pursue this area, what advice would you give?
    - What skills should I be working on in school?
    - How can I get more information?

KEY TO SELF-EXPRESSION
Discovery Through Journal Writing

To record your thoughts, use a separate journal or the lined pages at the end of the chapter.

Writing a journal requires a high level of reflection that goes way beyond a "Dear diary" approach. Reflection is an essential element of a scientific mind and critical thinking. The ability to observe yourself and your thoughts and feelings is a valuable step toward learning to observe the world around you. Thinking about your thoughts, feelings, and the events that occur each day will assist you in developing an observant mind as well as sharpen your imagination and creativity, which is required to understand and work in the sciences.

Start the journal process by writing a detailed description of your environment. You can go into the backyard, into the kitchen, or onto your front porch. Take as long as you need to do this exercise. Minimum: 10 minutes of continuous writing; maximum: several days, or weeks, if that helps you get all the details as precise as possible. (If you need help, consider the following question as a starting point: What do you see, smell, hear, feel?)

Name _____ Date _____

Journal

Journal

Name _____ Date _____

Discovering Science

Exploring Your Options

This chapter is about exploring the many opportunities and options you have by choosing the sciences as your field of study. In this chapter, you will learn what skills you already possess and what skills you need to develop to succeed in science. A number of career areas will be discussed so that you can begin to understand the pros and cons of different fields of study. The main goal of your exploration is to find something that not only interests you but fascinates you. Physicist John Trauger of the Jet Propulsion Laboratory puts it like this:

Scientists do what fascinates them, and what fascinates them is not something you can discover with science. They're interested in investigating where planets come from, say, not because science tells them to do that, but because as human beings they find that interesting. They go after questions they consider worth the investment of a lifetime.[1]

In this chapter, you will explore answers to the following questions:

- How can you find what fascinates you?
- What skills do you need to develop to succeed in science?
- What skills do you already have to succeed in science?
- What are the fields of study and careers in science?
- What are some of the barriers women face in technical sciences?
- How can you start thinking about choosing a specific science major?

HOW CAN YOU FIND WHAT FASCINATES YOU?

Since he was 5 years old, Stephen Jay Gould knew he wanted to be a paleontologist. Gould, now a Harvard professor, prolific science writer, and recipient of numerous awards, made his decision on his first visit to the dinosaur exhibit at the Museum of Natural History. He quickly became enthralled with a scientific field that seeks to unlock the mysteries of life through the study of fossils and ancient bones. And he continues studying intensively to this day.

On the other hand, not everyone is fortunate enough to discover what they want to do at such an early age. For instance, during Stephen W. Hawking's school years, nothing captured his undivided attention; yet he became what many consider to be the greatest theoretical scientist since Einstein. Hawking drifted from subject to subject, and although he had an immense interest in science, it wasn't until he was in his early twenties and diagnosed with a disabling disease, atrophic lateral sclerosis (ALS, or Lou Gehrig's disease), that he decided to buckle down, work hard, and really excel in one area. The area he chose was cosmology. Today, Hawking is a renowned scholar for his groundbreaking work on black holes and the origin of the universe.

Barbara McClintock spent years studying corn kernels to learn something about the inheritance of color. She wasn't even looking for what eventually made her famous: how genes can cross over to make genetic mutations that can cause cancer. Unfortunately, in 1950 her discovery did not go along with current genetic theory and she was ignored, or worse, thought of as unscientific. In the 1970s, when microbiologists noticed the same event in bacteria, McClintock was finally recognized, and in 1983 she won the Nobel prize for a discovery she had made thirty years earlier.

While reading this book, you may be in a position, like Hawking, of uncertainty—wondering what area of science interests you. Or you may be wondering if majoring in science will help you land a job when you graduate. These are common concerns, and ones many other college students, new and returning, are pondering. If, on the other hand, you are more like Stephen Jay Gould or Barbara McClintock and have already decided on your area of interest, you will pursue it with conviction no matter what the job market is like, or whether you are recognized or not. In any case, I congratulate you, because no matter which position you are in, certain, uncertain, or any place in between, a solid background in science will be extremely valuable to you.

"Tell me to what you pay attention and I will tell you who you are."
JOSÉ ORTEGA Y GASSETT

"When was I born? Where did I come from? Where am I going? What am I?"
THE HOPI QUESTIONS

Whether you end up with a career in science or not, the most important thing is that you are doing something in college that you like, are interested in, and, dare we say it, have a passion for.

Interest and passion in science does not mean you have to have a high GPA to be a science major, nor do you have to have a clear idea of what you want to do when you graduate. A great selling point of the educational system in the United States is that you get to keep trying. Just because you've had low scores in science in high school doesn't mean you can't succeed as a science major in college, providing you do two things: Love science and study with determination.

No matter what you do in the sciences do it because you enjoy and have an interest in science. You will succeed fueled by your interest, passion, and a great deal of determination.

WHAT SKILLS DO YOU NEED TO DEVELOP TO SUCCEED IN SCIENCE?

Creativity

Many scientific discoveries occur from using a creative mind or a mind that can see things just a little bit differently from others. Having a broad educational background in literature, philosophy, and politics will help you develop the ability to view problems in fresh new ways. Each discipline has its own methods for gaining knowledge and for understanding the world. The more of these disciplines you master, the better your ability to be flexible and adaptable.

Mathematics

You've heard people freely admit they have a problem with math, a "math block," or hatred of math, but have you ever heard anyone admit the same about reading? "I can't read." "I just can't seem to learn how to read." Admissions such as these are shameful, yet saying the same about math is acceptable. What does this tell you about our view of math? The ability to use quantitative skills, statistics, or in any way work with numbers, is essential to the information, or knowledge, age.

Grant Writing

The ability to find money to fund research and development includes the use of excellent and persuasive verbal and writing skills. The development of these skills occurs in English composition, debate, speech, and all courses requiring written papers. Use the opportunity college presents to work on these critical skills.

Diligence

Sticking with a project despite initial, or repeated, problems is especially important to the research scientist. Attention to detail and careful execution of instructions can be developed in all science lab classes and in all classes where you have to follow instructions to complete assignments.

Observation

Studying nature and statistical patterns, seeing things that others may not notice, or making new discoveries requires astute observation skills. All scientific endeavors demand the ability to see, hear, smell, and touch. Observation skills can be enhanced by taking a plant classification class where you learn to identify plants and trees. Suddenly, a whole new world is opened to you as you begin to notice attributes in plant life you never noticed before. The same is true of any scientific area from the night sky to a sick patient to microscopic organisms to DNA.

WHAT SKILLS DO YOU ALREADY HAVE TO SUCCEED IN SCIENCE?

Science Ability

If you love science and are good at it, you have it made. If you like, or love, science and are fair at it, you also have it made, although perhaps you will need to work harder. If you like, or love, science but have a hard time with it, or if it's been so long since you took a science class that you don't remember much about it anymore, don't give up. Two things will help you succeed: determination and a tutor. Determination is your job and a tutor is your school's. Free tutoring is available because almost all graduate students work in this role as part of their education and training.

Nancy Hoffman, nursing advisor at Washington State University, confided that when she decided to return to school as a science major after raising her two children, the thought of taking chemistry terrified her. All she could think of was how much she hated high school chemistry. When she returned to school, the first class she took was biology. Her grade was a C. She said, "I thanked God every day for that C."

She knew that if she was to continue in school with any success, she would have to find a way to get through chemistry. She explained, "Chemistry was like traveling to a foreign land where the people spoke a foreign language. I couldn't understand any of it." So she went to her school's learning center and found a tutor who was a graduate student in chemistry. For the first two months of the semester, she met with him after every class to review the material. Her final grade in organic chemistry? An A.

HOW TO FIND A TUTOR

This book will help you become a better student, which will go a long way toward ensuring your success as a science major, but part of succeeding is knowing when to ask for help. Tutors are an excellent source of help with any class. Contact your school's Academic Office, adult education center, or ask your advisor, teaching assistant, or instructor to help you find one. The people at your school want to help you succeed in college, and they will likely bend over backward to assist you. It's possible they are sitting in their offices right now, waiting for you to come see them.

Now she knew how to review on her own, or with other students, and received As in all her science classes. She advises all students to visit their school's learning center to find out who, or what, can help them. She explained, "Don't let embarrassment keep you from asking for help."[2]

Interest in Science

Science is part of everyone's daily life, but some notice it more than others. There are those who not only notice it but read about it, experiment, study, and watch every PBS *NOVA* episode they can. If you are one of these people, you understand that the variety of fields of study in science can appeal to anyone with the basic desire to study science. For instance, as pictures have been received from the *Hubble Telescope*, many people's interest in astronomy has been sparked, and for good reason. Looking at a picture of a cloud nebula from the *Hubble Telescope*, NASA administrator, Dan Goldin, said:

> You ask fifty questions after looking at that picture. That cloud on the left is six trillion miles high. That's one light-year. The first thing that comes to my mind is: How, as a species, did we attain the intellect that went into making the tool that could take that picture? Seven thousands light-years from Earth, and look at the clarity we got in that picture. That picture is creation. You're seeing stars being formed if we understand it right. When I showed this picture in Bozeman, Montana, the local merchants stood up and applauded.[3]

WHAT ARE THE FIELDS OF STUDY AND CAREERS IN SCIENCE?

Sciences are divided according to what you study rather than by how you study it. For example, *how you study* involves methods of scientific research. Some studies occur only in a laboratory; others rely strictly on research in the field. An example of field study is Charles Darwin's work on the development of life in the Galápagos Islands. An example of laboratory study is Barbara McClintock's groundbreaking discoveries in genetics of crossover.

The most familiar classification of sciences is by the area of study, that is, *what* you study. The main categories of these areas are:

BIOLOGICAL SCIENCE
- Aquatic biology
- Biochemistry
- Botany
- Physiology
- Zoology
- Ecology
- Agricultural science
- Health science

PHYSICAL SCIENCE
- Physics
- Astronomy
- Meteorology
- Chemistry
- Geology

MATHEMATICS
   Theoretical
   Applied

COMPUTER SCIENCE
   Computer engineering
   Systems analysis

ENGINEERING SCIENCE
Mechanical
Civil
Electrical
Physical
Chemical

Biological Science

What Is It?

Biology is the study of life. Life and its environments make up complex interacting systems called *ecosystems*. Biology also encompasses the history of life, as studied through fossils and evolution. As a student of biology, you will study single-cell organisms, cells that make up organisms, and entire organic life cycles and environments. The study of biology can take you to the rain forests of Brazil, the deserts of Africa, or to the world's deepest oceans. The work of biologists is achieved in laboratories using highly specialized and intricate technology or in rustic conditions using pencils, notebooks, and observation.

Biology is a complex field of study that necessitates a solid background in chemistry, physics, and math. The advantage of studying a multidisciplinary field like biology is that it will give you a flexible and versatile knowledge of the sciences, an important feature for success in science. Biology is an excellent background for health, agricultural, or environmental sciences.

Biology includes, but is not limited to, the study of:

- Cell function, structure, and chemistry
- Inheritance traits, chromosomes, and genes
- Evolution and changes through time
- Viruses, monerans, protists, and fungi
- Invertebrates: insects, sponges, starfish
- Vertebrates: fish, mammals, reptiles, birds
- Ecology of populations and communities
- Human structure and function, body framework, nutrition, sleep

What Careers Are There in Biology?

In biology, the health sciences are one of the fastest-growing career areas today. Positions can involve working with people and technology or performing research to improve health and fight diseases. Because of the variety of careers available in this field, I will discuss them separately later. Advances in the field of genetics has fueled extensive growth in biotechnology research. Genetic engineering, used to create new medicines and treatments for illnesses such as cancer, has opened up many new careers in research and development. The Human Genome Project, an international effort to map the entire human genome by the year 2005, has led to advances in the development of gene therapy and the prediction of future disease predisposition.

As advances in genetics take place, the field of ethics in science grows. Important questions arise regarding privacy, the handling of medical records, prolonging life, deciding who receives the benefit of new discoveries, and who pays the bills. New technology is creating the need for people trained not only in the technology but also in understanding the social and ethical implications of using that technology. Growth in biology careers is expected to be faster than average job growth from 1996 to 2006, but students should expect competition for research positions of any kind.[4]

Environmental biologist. Environmental issues have become increasingly important as the world population grows and resources dwindle. The environmental sciences are an area of biology that requires people trained to understand the complexities of managing natural resources, such as water, soil, and forests. Environmental impact studies are needed for almost every new structure being built or project, such as new roads, being planned today. Many government agencies, such as forest services and wildlife and fisheries departments hire those with bachelors, masters, and Ph.D.s in biology. Private industry also needs biologists for forestry, mining, fishing, and many other areas that impact life systems, which, when you think about it, covers almost everything.

Aquatic biologist. The study of plants and animals in water—limnologists in freshwater and marine biologists in saltwater—makes up aquatic biology. This is not to be confused with oceanography, which studies physical science, that is, the characteristics of the ocean.

Biochemist. Biochemists study the chemistry of living things. Along with molecular biologists, they often work using biotechnology to study reactions involved in metabolism, growth, and reproduction of living systems.

Botanist. Botanists study plant chemistry, structure, diseases, and environment. They also study the natural history of plants by examining geological, or fossil, records.

Microbiologist. Microbiology is the study of microscopic organisms such as fungi, bacteria, and algae. Health, or medical, microbiologists study disease organisms. Specialized areas include the environment, food, agriculture, virology (study of viruses), and immunology (study of how the body fights disease).

Physiologist. The functioning of plants and animals as whole organisms or at a cellular level is the field of physiology. Specialized areas of study include photosynthesis, movement, and reproduction.

Zoologist. Zoology is the study of living animals in their natural environment or of dead animals through dissection (to learn about structures). Zoology is usually divided by the type of animal studied, for example, mam-

> ### INVENTING YOUR OWN AREA OF BIOLOGICAL STUDY
>
> Plant chemist Eloy Rodriguez warns undergraduate students to avoid specializing in any one area too soon, and he should know. From his broad science background and his work in plant chemistry, he invented a whole new area of science called zoopharmacognosy. Zoopharmocognosy is the study of wild animal's possible use of plants for medicinal purposes.
>   Study a wide range of sciences, and then, if you can't find a career area that meets your needs, invent your own.[5]

malogists study mammals, herpetologists study reptiles, ornithologists study birds, and ichthyologists study fish.

Ecologist. Ecologists look at the big picture by studying organisms and their environment together. Population size, pollution, and geographic characteristics are also studied.

Agricultural scientist. Farm crops and animals are the interests of agricultural scientists who work in food, plant, soil, and animal science industries. Biotechnology and advances in genetics play significant roles in this area of study.

What Level of Education Is Needed for a Career in Biology?

A Ph.D. is needed for most independent research and university faculty positions. A master's degree is needed for the majority of research involving the development of products. A bachelor's degree is needed for nonresearch jobs.

Resources

Consult with your advisor or professors, or go to your local bookstore's science section where you will find a number of excellent books on all areas of biological science. Another useful source is the following organization:

American Institute of Biological Sciences
1444 I St. NW Suite 200
Washington, DC 20005
Home page: http://www.aibs.org

Health Science

What Is It?

Although the health sciences are part of the biological sciences, they are discussed separately to give you an idea of the careers and training available for a career in this rapidly growing field. The growth of health science is occurring for myriad reasons, for example:

- *The U.S. population is aging.* Everyone is not only getting older but they are living longer. This means more people who require health care and who want to remain healthy and active longer.
- *New technology.* Advances in technology have changed health care systems. With new techniques, surgical procedures that once required patients to stay a week in the hospital now require only day in an outpatient clinic. New drugs, new therapies, computers, and machines for testing and monitoring diseases have also changed health care and the need for well-educated professionals.
- *Healthy living.* The reason health care is beginning to focus on keeping people healthy is twofold. First, people are demanding it, and second, it saves money. Keeping people healthy saves the health care system millions of dollars, because it costs less to monitor and screen for problems than to treat to treat those same problems in an emergency room or hospital after they occur.

What Careers Are There in the Health Sciences?

The following job descriptions are adapted from the Washington State Healthcare Human Resources Association's "Descriptions of Occupations."

Biomedical equipment technician. Installs, maintains, repairs, calibrates, and modifies electronic, electrical, mechanical, hydraulic, and pneumatic instruments and equipment that are used in medical therapy, diagnosis, and research. May be involved in the operation, supervision, and control of equipment.

Dental assistant. Receives and prepares dental patients and assists the dentist during treatment. Tasks include chairside assistance to the dentist and may include maintaining and sterilizing instruments, taking and processing X-rays, and preparing dental compounds. May include some clerical duties.

Dental hygienist. Performs patient examination procedures, notes conditions of decay and deviation, administers local anesthetic, places temporary and permanent restorations, takes and develops X-rays, and educates patient about proper dental health habits. Works with the supervision of a dentist.

Dental laboratory technician. Creates dental prostheses, replacement for natural teeth, from the specifications of a dentist.

Dentist. Diagnoses and treats diseases of the teeth, gums, bone, and soft tissues of the mouth. Helps patients prevent dental problems through education and treatment.

Dietitian. Provides expertise on food and nutrition in relation to health; sometimes called a nutritionist. May work in a variety of settings including medical facilities. **Dietetic technologist** works under the supervision of a dietitian in handling routine duties of the food service team.

Electrocardiograph (ECG) technician. Operates a specialized machine to obtain readings of the electrical activities of a patient's heart.

Electroencephalogram technician (EEG). Operates instruments that measure and record electrical activities of the brain. Records are used to diagnose and assess disorders of the brain.

Emergency medical technician (EMT). Provides immediate emergency care to critically ill and injured patients and may drive the ambulance. Determines the nature and extent of the illness or injury and establishes priorities for emergency care.

Health services administrator. Manages hospitals or other health care facilities and their staff to assure good patient care. Duties often include developing and administering policy and programs, budgeting, planning new facilities, marketing, public relations and personnel management.

Licensed practical nurse (LPN). Works under the direct supervision of physicians or RNs to administer medications, monitor equipment, maintain patient charts, and take vital signs. May work as a private duty nurse in home or hospital.

Medical assistant. Performs office, laboratory, and treatment functions to assist physician in caring for patients. Work is generally "front office" (administrative tasks of reception/secretarial, medical machine transcription, bookkeeping, and medical insurance) or "back office" (sterilization of equipment, supply readiness, and may include taking patient's history, blood pressure, temperature, respiration, and pulse, and administering routine tests).

Medical laboratory technician/Medical technologist. Performs complex chemical analyses and routine chemical and biological tests on blood and body fluids using a variety of laboratory techniques and instruments; these tests are used in the prevention, detection, and diagnosis of disease.

Medical social worker. Helps patients and their families with problems that accompany illness or inhibit recovery and rehabilitation. Collects patient information to assist other health professionals to understand the social, emotional, and environmental issues underlying a patient's illness. Provides emotional support, makes referral to other helping agencies, and often makes discharge plans.

Nurse practitioner. Has additional nursing training. May perform physical examinations and diagnostic tests, develop and carry out treatment plans, and counsel patients about their health. Some work in specialty areas and some work independently.

Nursing assistant certified (NAC or CNA). Provides patient care with the direction of a nurse. Duties often include daily patient hygiene; feeding; admitting and discharging; taking and charting vital signs; may work in a hospital, nursing home, or doing home care.

Occupational therapist. Plans and organizes activities to improve the functioning of patients who are physically, mentally, or emotionally disabled. Plans and directs educational, vocational, and recreational activities to help patients become sufficient in self-care and activities of daily living.

Optician. Prepares and dispenses contact lenses, eyeglasses, and other eyewear according to the written prescription of an ophthalmologist (M.D.) or optometrist (D.O.).

Ophthalmic technologist. Assists ophthalmologist by using sophisticated equipment/techniques to gather information during eye exams. Also, assists with eye surgery, using microscopic and intricate instruments. Instructs patients about diagnosis/treatment.

Orthotist/Prosthetist. Designs, writes specifications for, and fits artificial appliances for disorders, following the prescription of a physician. Appliances include artificial arms and legs, and back braces, and surgical supports. **Orthotic technician** makes and repairs orthotic devices such as surgical corsets and corrective shoes. **Orthotic assistant** provides care to patients with disabling conditions of the limbs and spine. **Prosthetics technician** makes and finishes artificial limbs. **Prosthetic assistant** provides care to patients with partial or total loss of limbs.

Pharmacist. Provides drug information to physicians, nurses and other health care practitioners to assure optimal uses of drugs in patient care. Prepares compounds and dispenses medicine. Provides information and educates consumers about medicines—their storage and use. Works with pharmaceutical industry to develop and sell new drug products.

Physical therapist. Evaluates, plans, and administers treatment to patients with problems related to muscular and skeletal systems. Long- and short-term problems are generally the result of injury, disease, or surgery. Administers and interprets tests for muscle strength, coordination, range of motion, respiratory, and circulatory efficiency. Educates patients and their families about the use and care of treatment equipment.

Physician. Diagnoses and treats human disease with medicines and other treatment. Works to maintain and improve the health of their patients. Some specialize in one field of practice, others teach or do research.

Physician assistant. Is a highly skilled professional who under the supervision of a licensed physician, performs many of the routine duties usually done by a doctor. (Washington State allows independent practice with the coordination of a physician.)

Radiologic technologist. Works with a diverse group of professionals to perform a variety of diagnostic studies using ionizing radiation. X-ray technicians use radiation to make images of the internal organs and bones of the body; develop x-ray film; prepare records of findings; and make minor adjustment to equipment. **Nuclear medicine technician** uses radioactive materials to form images of organ systems to determine the presence of disease. **Diagnostic ultrasound technician** creates diagnostic images using high-frequency sound. **Radiation therapy technician** uses various forms of radiation to treat cancer and other diseases. Other technician fields include **MRI, CT,** and **special procedures;** each uses specialized equipment to diagnose and treat disease.

Recreational therapist. Is a specialist within the recreation profession and is involved in organizing, administering, and presenting therapeutic recreational activities for patients that make a definite contribution to the recovery of or adjustment to an illness, disability, or specific social problem.

Registered nurse. Works to promote health and wellness and prevent disease using substantial, specialized knowledge, judgment, and skill based on the principles of the biological, physiological, behavioral, and sociological sciences. Increasingly, nurses specialize in one area of practice, provide education and training, or work in the administration of health care programs.

Rehabilitation counselor. Assists clients with physical, mental, emotional, and social disabilities to adjust to their disabilities and find suitable employment. Develops a service plan that may call for occupational training, medical therapy, psychotherapy, or counseling. Works with employers to redesign jobs to fit the capabilities of the client.

Respiratory therapist. Treats patients with heart/lung disorders under the direction of a physician. May give temporary relief to patients with chronic breathing problems or emergent problems of heart failure, stroke, drowning, and shock. May instruct patients on the use of equipment and conduct educational sessions on disease treatment and prevention.

Speech pathologist and audiologist. Specializes in communication disorders. A speech-language pathologist is primarily concerned with language problems, while an audiologist is concerned with hearing disorders. These specialists test, evaluate, and plan therapy to restore as much normal functioning as possible. Works closely with teachers, rehabilitation counselors, and medical professionals.

Surgery technician. Shares the responsibilities of the operating room team before, during, and after surgery. Assists surgery team scrub; arranges sterile instruments; prepares the patient; passes instruments to the surgeon and the assistants; maintains supplies; transports the patient after surgery; and performs related duties.

Veterinarian D.V.M./veterinarian technician. Veterinarians, assisted by technicians, treat and prevent illnesses in pets, livestock, and marine animals.

What Level of Education Is Needed for a Career in the Health Sciences?

Education requirements vary depending on the field you wish to pursue. With a two-year degree, you can work as a technician; with a bachelor's degree, you have more options; and, of course, with a graduate degree even more. A Ph.D. is needed to teach at the university level. In general, the more education you have, the more autonomous, or independent, you can be in your field.

Resources

J. R. Katz (1999). *Majoring in Nursing: From Prerequisites to Postgraduate and Beyond.* (New York: Farrar, Straus & Giroux, 1999). An in-depth look at the health science's largest profession, nursing.

For links to many health career information sites, see:

http://www.nursingworld.org

http://www.nih.gov

Physical Science

What Is It?

Physical science, a large and varied field of study, is concerned with the world of energy and matter. Physical scientists study how things work on the earth and in the rest of the universe. Areas of study are classified by the type of energy or matter studied and include:

- Physics and astronomy
- Meteorology and climatology
- Chemistry
- Geology
- Geophysics

Job growth in the areas of chemistry and geology is expected to be average, but in physics and meteorology, slower than average.[6]

What Careers Are There in Physical Science?

Physicist. The principles, structure, and interactions between energy and matter are what physics is all about. The knowledge gained from physics theory is applied in the development of materials, electronics, medical technology, and medical equipment. Specialized areas in physics include nuclear, particle, atomic and molecular, condensed matter, chemical, biophysics, optics, acoustics, and fluids.

Astronomist. Astronomers study the universe, including the sun, planets, stars, and galaxies. Knowledge from astronomy is applied to space flight and into developing the technology that collects cosmological data. Most astronomers are involved in research.

Meteorologist. The study of the earth's atmosphere's physical characteristics, movement, and how it affects the environment is the realm of meteorologists. Many meteorologists study the weather. Research is applied to air pollution control, global warming, and ozone depletion. Climatology is a specialized area of meteorology that studies rainfall, temperature, and wind in relation to agriculture and land use.

Chemist. Chemistry includes everything in the living and nonliving world. Chemistry research is applied to drugs, fibers, plants, energy, pollution, and agriculture. Analytical chemists study the composition and structure of substances; organic chemists study carbon-based chemicals and their use in the synthesis of thousands of products; inorganic chemists study non-carbon-based chemicals; and physical chemists study molecular and atomic reactions.

Geologist. Geology is the study of the physical and historic nature of the earth. Geologists study rocks, earthquakes, oil, natural gas, minerals, and groundwater. The history of the earth is studied through the fossil records and sedimentation. Specialized areas of study include:

>   Paleontologists (fossils and evolution)
>   Mineralogists (precious stones and minerals)
>   Oceanographers (ocean floor)
>   Petroleum geologists (oil and natural gas)
>   Seismologists (earthquakes and faults)
>   Volcanologists and geochemists (igneous and metamorphic rock)
>   Hydrologists (underground and surface water and precipitation)

What Level of Education Is Needed for a Career in Physical Science?

A Ph.D. is required for most physics, astronomy, and geology work because of the emphasis on basic research (as opposed to applied research) in those fields. Master's degrees are useful for applied research and development or teaching in a two-year community college. Those with bachelor's degrees will likely work as technicians or in ecological areas such as waste management.

In meteorology a bachelor's degree is a minimum requirement for most government and private industry positions. The American Meteorological Society offers certification if applicants meet their educational requirements and pass an examination.

Undergraduate chemistry majors usually take math, physics, biology, and chemistry courses. Many chemists also have a background in business or economics. A bachelor's degree is usually required for entry-level positions in chemistry.

Resources

American Geological Institute
Home page: http://www.agiweb.org

Geological Society of America
Home page: http://www.goesociety.org

American Chemical Society
Education Division
1155 16th St. NW
Washington, DC 20036 American Meteorological Society
Home page: http://www.ametsoc.org/AMS

American Institute of Physics, Career Planning and Placement
Home page: http://www.aip.org

Mathematics

What Is It?

Mathematics is the oldest science, and it is used today to create new theories that explain technology, economics, engineering, and business-related problems. Mathematical models can be used to predict changes in populations, marketing trends, or encryption codes. Applied mathematics is used in a variety of careers because math is the foundation of many sciences, including psychology and social science. Careers that use a great deal of math are actuaries, operations research analysts, and statisticians.

What Level of Education Is Needed for a Career in Mathematics?

A bachelor's degree is the minimum degree necessary for a career in math. In many government jobs, even entry-level jobs, a bachelor's degree with a math major is required. In industry, a bachelor's degree will be needed but the job titles may be computer programming, analyzing, or engineering. A dual degree in math and another area, such as teacher education, computer science, engineering, and geology or other physical science, is a plus in the job market. A Ph.D. or master's degree is needed for research and development and university teaching.

Resources

Conference Board of the Mathematical Sciences
1529 18th St. NW
Washington, DC 20036

Computer Science

What Is It?

Computer science is expected to be one of the top three fastest-growing occupations.[7] As computers are becoming rapidly integrated into daily life and routinely used in all areas of the economy, highly trained people to design hardware and software and to convert outdated systems as new technology becomes available will be needed. A background in other areas of science, such as math and business, can also be important to a computer science career. The opportunity for self-employment in computer science is available for creative and innovative entrepreneurs.

What Careers Are There in Computer Science?

Computer scientist. Computer scientists design computers and software and therefore must have the ability to solve complex problems and make creative applications of theories. Specialized areas include computer theorist, researcher, and inventor; jobs are also available in academic, private industry, and small business settings.

Computer engineer. Mainly concerned with the hardware and software of computers, a computer engineer applies many of the theories generated by computer scientists to the designing and building of systems. Specialized areas include hardware and software engineers and developers.

Systems analyst. Solving computer problems and finding design solutions in the workplace is the big challenge of systems analysts. They work with others in an organization to analyze problems, set goals, design steps to meet the goals, and determine what hardware and software is needed. Many systems analysts work on networking computers to allow different types of systems to communicate with each other.

Other areas of specialization include database administrators, security and telecommunications specialists, and a growing number of Internet and World Wide Web positions.

What Level of Education Is Needed for a Career in Computer Science?

Computer science is similar to the other areas of science in that a Ph.D., or at least a master's degree, is needed to teach or do research for industry or universities. A bachelor's degree is regarded as essential for entry-level positions with experience coming in a close second. It is possible to work as a computer scientist with a different science major as long as you have strong math, logic, and problem analysis skills.

Certification is required by some employers and is available through the Institute for Certification of Computing Professionals. Requirements include four years' work experience, or two years' experience and a college degree, and, of course, passing an examination.

Resources

Association for Computing
1515 Broadway
New York, NY 10036

Computer Society
1730 Massachusetts Ave. NW
Washington, DC 20036-1992
Home page: Institute for Certification of Computing Professionals
http://www.iccp.org

Engineering Science

What Is It?

"An engineer is a problem solver who uses technical means to make theory a reality."[8] Engineer and writer John Garcia defines engineering as solving problems using theoretical tools based on mathematics and scientific observations. Engineering has a lengthy history that goes back to ancient civilizations where engineering principles were used to build complex structures, such as the pyramids of Egypt, roads, and waterways for crop irrigation. At the turn of the nineteenth century, science experienced significant advances due to electricity and later in the twentieth century, with computers and a multitude of new materials. Likewise, the field of engineering has grown with each new advance in science and technology.

What Careers Are There in Engineering Science?

The main areas of engineering are civil, mechanical, chemical, and electrical. Other specialty areas include mining, metallurgy, computers, nuclear science, industrial systems, agriculture, biomedicine, ceramics, geology, and manufacturing. The largest number of engineers are electrical, followed by mechanical, civil, industrial, aerospace, and chemical.

Civil engineer. Civil engineers design and build structures such as roads, bridges, waterways, dams, buildings, and sewage treatment plants.

Mechanical engineer. "Mechanical engineering is basically the study of civil engineering systems that move," says Keri Sobolik, a mechanical engineer.[9] The design of mechanics, or machines, that transmit, produce, or use energy is part of mechanical engineering. Included with this career are the newer fields of robotics and computer-aided design, or CAD.

Chemical engineer. Stephanie Witkowski defines her field as, "a cross between physical chemistry and mechanical engineering."[10] A new area in the field is working with genetically engineered products such as artificial muscle tissue for people with muscle diseases or injuries.

Electrical engineer. Electronics play a huge part in our economy and daily life. The Bureau of Labor Statistics show that the majority of engineers graduate with degrees in electrical engineering.[11] These engineers work on electrical circuitry, electromagnetics, control systems, and signal processing. A newer specialty is the development of neural networks using the human brain's neurons as models for programming systems—an area that combines an interest in life sciences with engineering.

What Level of Education Is Needed for a Career in Engineering Science?

A bachelor's degree in engineering is necessary for entry-level positions, and with that degree you will have excellent opportunities to specialize in the area you choose. Graduate degrees are needed for academic teaching or research.

Resources

J. Garcia (1995). *Majoring in Engineering: How to Get From Your Freshman Year to Your First Job* (New York: Noonday Press, 1995). Excellent resource; the appendix of this book lists professional organizations, gender and cultural organizations, and student honor societies.

Alternative Careers in Science

There are many other careers in science besides the ones mentioned in this chapter. They include a career as a science writer, technical writer, broadcast science journalism, and science-based investment advisor. To discover more alternative career options in science, read Cynthia Robbins-Roth's book, *Alternative Careers in Science: Leaving the Ivory Tower.*

WHAT ARE SOME OF THE BARRIERS WOMEN FACE IN THE TECHNICAL SCIENCES?

To give you an idea of some barriers women in technical science fields, such as engineering, may face, the National Science Foundation has listed common problems. Here is a summary of their research. For more information check the web site below.

The following is adapted from the National Science Foundation.[12]

The Barriers

The work culture in engineering jobs was established by men. Women are newcomers who have to find a way in which they can keep their own identity and be accepted by their colleagues as competent engineers.

Hidden barriers are often based on the following assumptions:

"She will not be able to do that; I'll help her."

"She will not be interested in that kind of job; I will not ask her."

"Of course they will ask a man to do that; a man is better in that kind of thing."

The following arguments are often used against women engineers:

- Cannot function in a man's world
- Cannot make hard decisions where necessary
- Are not good at abstract thinking
- Will not be accepted by subordinates or by others
- Are too emotional

Women engineers frequently report differences between themselves and male engineers in the following:

- The way they approach problems

> # REAL WORLD PERSPECTIVE
>
> **Janis M. Wignall, B.S.—Science and Education Liaison,
> Immunex Corporation, Seattle, Washington**
>
> Janis M. Wignall has a degree in microbiology and virology and has worked for Immunex for fourteen years both as a bench scientist and, more recently, as a teacher. Janis was fortunate in her early education in England because she went to an all-girls school. Lucky, Janis says, because studies show that girls in grade school often don't succeed in science because they need more "wait time"; girls need longer to answer questions than boys do. In middle school, she notes, it is generally considered "not cool" to be a girl and show too much interest or ability in science. Now, girls and young women interested in science can find support in many organizations formed by women professionals, especially for the purpose of increasing the number of women in science (see appendix for resources).
>
> Janis claims two reasons for her success in science: her family background—her mother worked in science—and her learning style. "Being a kinesthetic learner helps in doing lab work. I was hopeless in exams, but with hard work and my lab projects, I did fine," she says. Janis participated in many lab projects as an undergraduate and attributes her two job offers upon graduation to this work. She emphasizes the need for science students to begin immediately finding projects to obtain hands-on experience. Checking with your professors or seeking part-time work in a science lab are two methods. For students that have to work while in school, she suggests trying to find work in a science situation. "Most students have ideas about what they want to do," she says, "but working in different areas will help make a decision and narrow down interests."
>
> Another way to gain experience is by taking electives in independent research. This allows you to focus on one area, practice your critical thinking and problem solving skills, and gain important experience for your resume. This type of resume, according to Janis, gets students in the door of Immunex, along with stellar "soft" skills like the ability to communicate, work

- The way they approach people
- The content of the "small talk" at work (cars, soccer, women)
- Behavior in general, such as avoidance of more personal topics, kind of humor, way they present themselves, etc.

Women engineers say they do not differ from male engineers in:

- The ability to make hard decisions
- Abstract reasoning

The Solutions

Women engineers consider the following solution to be most effective for coping with the barriers they confront:

- Perform well; make yourself accepted by simply "being good."
- Be a good organizer.

- Learn to be more assertive.
- Learn to recognize the little tricks that are played under the surface.
- Be very explicit about what you want and what you do not want.
- Try to understand why men act as they do; explain to them where their assumptions go wrong.
- Learn how to negotiate effectively.
- Stay your real self; be self-confident, flexible, and creative in finding new solutions for old and new problems (engineers like each new problem that comes their way!).
- Keep your sense of humor.
- Never leave the labor market (there is no way back).
- Use mentorship, and when available, use female mentors.
- Encourage women networks within the company.

Men in Engineering Today

The data on young male engineers shows that a growing number of them no longer have aspirations to reach the top and to sacrifice everything to get there. They want interesting and challenging work. In addition, male engineers want to have time available for other things in life besides work. For many of them this means taking an active part in the upbringing of their children, but it can also be for spending time on other interests besides work.

HOW CAN YOU START THINKING ABOUT CHOOSING A SPECIFIC SCIENCE MAJOR?

Major
A subject of academic knowledge chosen as a field of specialization, requiring a specific course of study.

While many students come to college knowing what they want to study, many do not. That's completely normal. College is a perfect time to begin exploring your different interests. In the process, you may discover talents and strengths you never realized you had. For example, taking an environmental class may teach you that you have a passion for finding solutions to pollution problems.

While some of your explorations may take you down paths that don't resonate with your personality and interests, each experience will help to clarify who you really are and what you want to do with your life. Thinking about choosing a **major** involves exploring potential majors, being open to changing majors, and linking majors to career areas.

Exploring Potential Majors

Here are some steps to help you explore majors that may interest you.

Take a variety of classes. Although you will generally have core requirements to fulfill, use your electives to branch out. Try to take at least one class in each area that sparks your interest.

Don't rule out subject areas that aren't classified as "safe." Friends or parents may have warned you against pursuing certain careers, encouraging you to stay with "safe" careers that pay well. Even though financial stability is important, following your heart's dreams and desires is equally important. Choosing between the "safe" path and the path of the heart can be challenging. Only you can decide which is the best for you.

Spend time getting to know yourself, your interests, and your abilities. The more you know about yourself, the more ability you will have to focus on areas that make the most of who you are and what you can do. Pay close attention to which areas inspire you to greater heights and which areas seem to weaken your initiative.

Work closely with your advisor. Begin discussing your major early on with your advisor, even if you don't intend to declare a major right away. For any given major, your advisor may be able to tell you about both the corresponding department at your school and the possibilities in related career areas. You may also discuss with your advisor the possibility of a double major (completing the requirements for two different majors) or designing your own major, if your school offers an opportunity to do so.

Changing Majors

Some people may change their minds several times before honing in on a major that fits. Although this may add to the time you spend in college, being happy with your decision is important. For example, a student majoring in science education may begin student teaching only to discover that he really doesn't feel comfortable in front of students.

If this happens to you, don't be discouraged. You're certainly not alone. Changing a major is much like changing a job. Skills and experiences from one job will assist you in your next position, and some of the courses from your first major may apply—or even transfer as credits—to your next major. Talk with your academic advisor about any desire to change majors. Sometimes an advisor can speak to department heads in order to get the maximum number of credits transferred to your new major.

Whatever you decide, realize that you do have the right to change your mind. Continual self-discovery is part of the journey. No matter how many detours you make, each interesting class you take along the way helps to point you toward a major that feels like home.

Linking Majors to Career Areas

The point of declaring and pursuing a major is to help you reach a significant level of knowledge in one science area, often in preparation for a particular career. Before you discard a major as not practical enough, consider where it might be able to take you. Thinking through the possibilities may open doors that you never knew existed. Besides finding an exciting path, you may discover something highly marketable and beneficial to humankind as well.

For each major there are many career options that aren't obvious right away. For example, a student working toward a teaching certification in science doesn't have to teach public school. This student could develop curricula, act as a consultant for businesses, develop an on-line education service, teach overseas for the Peace Corps, or create a public television program. The sky's the limit.

Explore the educational requirements of any science career that interests you. Your choice of major may be more or less crucial depending on the career area. For example, pursuing a career in medicine almost always requires a major in some area of the biological sciences. Many employers are more interested in your ability to think than in your specific knowledge, and therefore they may not pay as much attention to your major as they do to your critical-thinking skills.

Joie de Vivre

The French have a phrase that has become commonly used in the English language as well: *joie de vivre*, which literally means "joy of living." A person with *joie de vivre* is one who finds joy and optimism in all parts of life, who is able to enjoy life's pleasures and find something positive in its struggles. Without experiencing difficult and sometimes painful challenges, people might have a hard time recognizing and experiencing happiness and satisfaction.

Think of this concept as you examine your level of personal wellness. If your focus on what is positive about yourself, that attitude can affect all other areas of your life. Give yourself the gift of self-respect so that you can nourish your body and mind every day, in every situation. Through both stressful obstacles and happy successes, you can find the joy of living.

Chapter 2   Applications

Name _____   Date _____

KEY INTO YOUR LIFE
Opportunities to Apply What You Learn

2.1 Learning From Others

TOOLS TO USE:

College catalog
Other students
Instructors and professors
Career center
Student organization members
Faculty Web sites

Using any or all of the tools above, chose one person who works in a scientific area of interest to you. Base your decision on the recommendations of others or from information from the faculty Web pages. This person should have work and research experience. They should be available to you, that is, have on-campus office hours. Call them and request a meeting.

Ask them the following questions and add any of your own:

1. What is your educational background? _____
_____

2. How did you decide on this area? _____
_____
_____

3. Was it hard to find work in the area and how did you go about it? _____
_____
_____

4. What areas are there that are related to this work? _____
_____
_____

5. Do you ever have students volunteer to work with you? _____

6. Can I contact you later if I have more questions? _____

Alertness to Real World Scientific Research in the News

To be a success in science, you must study all aspects of it including what is in the news. Pay attention to what appears in the popular press (newspapers, magazines, television news) and in scientific journals.

1. For one week review a local newspaper and the local television news. Keep notes on the science news presented.
2. In class, divide into small groups and discuss your notes, looking for common threads between the newsworthy information and other social or political trends.
3. In small groups, discuss whether there is a great deal of information about research on the prevention of air crashes just after a major airline accident occurs. What other news events spark science reporting?

Alertness to Real World Scientific Research in the Movies

Using recent science research in the movies is nothing new. When the Ebola virus was discovered, movies about viral epidemics were made. When astronomists reported their findings from meteor studies, movies were made about the end of the earth via a meteor collision.

1. List a movie that you think used a scientific research study or discovery for its main theme.

   _____

2. Find at least one article in a *scientific journal* that discusses this recent research discovery and list it here.

   _____
   _____

3. In class or in small groups, discuss how the movie did or did not distort the research or discovery. What effect does a movie of this kind have on the public's perception of science? Does it further it or hinder it?

   _____
   _____

2.4 Interests, Majors, and Careers

Start by listing activities and subjects you like.

1. _____
2. _____
3. _____
4. _____
5. _____
6. _____

Name three majors that might relate to your interests and help you achieve your career goals.

1. _____
2. _____
3. _____

For each major, name a corresponding career area you may want to explore.

1. _____
2. _____
3. _____

KEY TO SELF-EXPRESSION
Discovery Through Journal Writing

To record your thoughts, use a separate journal or the lined page at the end of the chapter.

Observing What You Already Observed

Return to your first journal entry and read it through. Next, return to the initial observation site and begin observing it again. Record all new observations in your journal. You should spend a minimum of 10 minutes of continuous, uninterrupted writing. When you are finished, read what you wrote. Think about how your first and second observations differed and how they where the same. Reflect on this observations process and write your thoughts, or feelings about it.

Journal

Name _____ Date _____

Self-Awareness

Knowing Who You Are and How You Learn

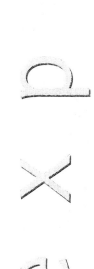

Learning is not something you do just in college. Throughout your life, learning can help you keep up with the rapid pace at which the world is changing.

Technology, for example, is changing so fast that you cannot learn today about all of the computer operations that will be commonplace five years from now. However, you can learn how to be an effective learner in school and in the workplace so that you can keep pace with changes as they occur. In this chapter, you will learn what recent studies show about the math and science abilities of students in the United States and how you can learn to be a better science learner. You will also become aware of your learning style by completing three different learning style assessments. Each assessment will add a different dimension to the picture you are forming of yourself. You will then explore other important elements of self: your self-perception, your preferences, your habits, your abilities, and your attitudes.

In this chapter, you will explore answers to the following questions:

- How well did your school prepare you for college science?
- Is there one best way to learn?
- How can you discover your learning styles?
- What are the benefits of knowing your learning styles?
- How do you explore who you are?

HOW WELL DID YOUR SCHOOL PREPARE YOU FOR COLLEGE SCIENCE?

In recent years, you may have seen headlines like *Newsweek's*, "An 'F' in World Competition," that emphasize the poor science and math abilities of U. S. students. The media have focused on test scores, especially in relation to other countries, but those scores do not necessarily represent the entire picture. The United States has students in many states with high scores when compared to other nations of the world, despite the fact that, as a whole, the United States falls behind other countries.[1]

Secretary of Education Richard W. Riley made the following points about high school education at a recent conference of mathematicians and math educators[2]:

- In the United States, math and science scores increased significantly from 1990 to 1996.
- U. S. students scored among the lowest in the world in mathematics and science, general knowledge; physics; and advanced mathematics.
- U. S. fourth graders scored as well as or better than fourth graders in many other countries did in math and science.
- U. S. eighth graders showed a decline in science and math performance.
- U. S. twelfth graders performed below the international average of the 21 countries tested.

There has been controversy among educators about the interpretation of international testing scores, but everyone seems to agree that U. S. schools need to be graduating students who are far better prepared in math and science than is currently the case.

Within the United States, there is great variation among high school students' scores, often depending on the state they live in. If you come from a state like Iowa or Utah, your high school is doing very well in preparing you in math and science. Many states performed as well as the students from the highest-scoring countries, according to international studies. Or you may live in a state where students had poorer scores; if so, your high school may not have prepared you to meet the challenge of college courses. You will have some catching up to do, but that's not an impossible feat. Students have been attending college and graduating with science honors from every state in the union. And you can, too.

IS THERE ONE BEST WAY TO LEARN?

Your mind is the most powerful tool you will ever possess. You are accomplished at many skills and can process all kinds of information. However, when you have trouble accomplishing a particular task, you may become convinced that you can't learn how to do anything new. You may feel that those who can do what you can't have the "right" kind of ability. Not only is this perception incorrect, it can also damage your belief in yourself.

There is no one "best" way to learn. Instead, there are many different learning styles, and different styles are suited to different situations. Your individual learning profile is made up of a combination of learning styles. Each person's profile is unique. Just like personality traits, learning styles are part of your personal characteristics. Knowing how you learn is one of the first steps in discovering who you are.

Discovering your favorite learning style will help you to develop study plans that work best for you, but you must learn to use all the learning styles. The next section will help you find out your preferred style, but that style will not be the only one you will need to succeed in the sciences. To be successful in science, you must become adept at all methods of learning.

Learning style
A particular way in which the mind receives and processes information.

HOW CAN YOU DISCOVER YOUR LEARNING STYLES?

Your brain is so complex that one inventory cannot give you all the information you need to maximize your learning skills. You will learn about and complete three assessments: the *Learning Styles Inventory*, the *Pathways to Learning* inventory based on the Multiple Intelligences Theory, and the *Personality Spectrum*. Each of these assessments evaluates your mind's abilities in a different way, although they often have related ideas. Your results will combine to form your learning styles profile, consisting of the styles and types that best fit the ways that you learn and interact with others. After you complete the various learning styles inventories, you will read about strategies that can help you make the most of particular styles and types, both in school and beyond. Your learning styles profile will help you to improve your understanding of yourself, how you learn, and how you may function as a learner in the workplace.

"To be what we are, and to become what we are becoming, is the only end of life."
ROBERT LOUIS STEVENSON

Learning Styles Inventory

One of the first instruments to measure psychological types, the Myers-Briggs Type Indicator (MBTI), was designed by Katharine Briggs and her daughter, Isabel Briggs Myers. Later David Keirsey and Marilyn Bates combined the sixteen Myers-Briggs types into four temperaments. Barbara Soloman, Associate Director of the University Undesignated Student Program at North Carolina State University, has developed the following learning styles inventory based on these theories and on her work with thousands of students.[3]

"Students learn in many ways," says Professor Soloman. "Mismatches often exist between common learning styles and standard teaching styles. Therefore, students often do poorly and get discouraged. Some students doubt themselves and doubt their ability to succeed in the curriculum of their choice. Some settle for low grades and even leave school. If students understand how they learn most effectively, they can tailor their studying to their own needs."

"Learning effectively" and "tailoring studying to your own needs" mean choosing study techniques that help you learn. For example, if a student responds more to visual images than to words, he or she may want to construct notes in a more visual way. Or, if a student learns better when talking to people than when studying alone, he or she may want to study primarily in pairs or groups.

Science courses use a variety of teaching methods. For instance, in one course you may be doing a great deal of reading, while in another you will do hands-on work in a laboratory or in the field. Most instructors use a variety of teaching methods to appeal to their students' different learning styles. Variety usually means good teaching because it helps keep your attention and also helps you learn other styles by practicing them.

The Dimensions of Learning

The Learning Styles Inventory has four "dimensions," within each of which are two opposing styles. At the end of the inventory, you will have two scores in each of the four dimensions. The difference between your two scores in any dimension tells you which of the two styles in that dimension is dominant for you. A few people will score right in between the two styles, indicating that they have fairly equal parts of both styles. Following are brief descriptions of the four dimensions. You will learn more about them in the section on study strategies, after you complete all three assessments.

Active/Reflective. *Active* learners learn best by experiencing knowledge through their own actions. *Reflective* learners understand information best when they have had time to reflect on it on their own.

Factual/Theoretical. *Factual* learners learn best through specific facts, data, and detailed experimentation. *Theoretical* learners are more comfortable with big-picture ideas, symbols, and new concepts.

Visual/Verbal. *Visual* learners remember best what they see: diagrams, flowcharts, time lines, films, and demonstrations. *Verbal* learners gain the most learning from reading, hearing spoken words, participating in discussion, and explaining things to others.

Linear/Holistic. *Linear* learners find it easiest to learn material presented step by step in a logical, ordered progression. *Holistic* learners progress in fits and starts, perhaps feeling lost for a while, but eventually seeing the big picture in a clear and creative way.

TERMS

Holistic Relating to the wholes of complete systems rather than the analysis of parts.

Please complete this inventory by circling **a** or **b** to indicate your answer to each question. Answer every question and choose only one answer for each question. If both answers seem to apply to you, choose the answer that applies more often.

1. I study best
   a. in a study group.
   b. alone or with a partner.

2. I would rather be considered
   a. realistic.
   b. imaginative.

3. When I recall what I did yesterday, I am most likely to think in terms of
   a. pictures/images.
   b. words/verbal descriptions.

4. I usually think new material is
   a. easier at the beginning and then harder as it becomes more complicated.
   b. often confusing at the beginning but easier as I start to understand what the whole subject is about.

5. When given a new activity to learn, I would rather first
   a. try it out.
   b. think about how I'm going to do it.

6. If I were an instructor, I would rather teach a course
   a. that deals with real-life situations and what to do about them.
   b. that deals with ideas and encourages students to think about them.

7. I prefer to receive new information in the form of
   a. pictures, diagrams, graphs, or maps.
   b. written directions or verbal information.

8. I learn
   a. at a fairly regular pace. If I study hard, I'll "get it" and then move on.
   b. in fits and starts. I might be totally confused and then suddenly it all "clicks."

9. I understand something better after
   a. I attempt to do it myself.
   b. I give myself time to think about how it works.

10. I find it easier
    a. to learn facts.
    b. to learn ideas/concepts.

11. In a book with lots of pictures and charts, I am likely to
    a. look over the pictures and charts carefully.
    b. focus on the written text.

12. It's easier for me to memorize facts from
    a. a list.
    b. a whole story/essay with the facts embedded in it.

13. I will more easily remember
    a. something I have done myself.
    b. something I have thought or read about.

14. I am usually
    a. aware of my surroundings. I remember people and places and usually recall where I put things.
    b. unaware of my surroundings. I forget people and places. I frequently misplace things.

15. I like instructors
    a. who put a lot of diagrams on the board.
    b. who spend a lot of time explaining. ✓
16. Once I understand
    a. all the parts, I understand the whole thing. ✓
    b. the whole thing, I see how the parts fit.
17. When I am learning something new, I would rather
    a. talk about it. ✓
    b. think about it.
18. I am good at
    a. being careful about the details of my work. ✓
    b. having creative ideas about how to do my work.
19. I remember best
    a. what I see. ✓
    b. what I hear.
20. When I solve problems that involve some math, I usually
    a. work my way to the solutions one step at a time. ✓
    b. see the solutions but then have to struggle to figure out the steps to get to them.
21. In a lecture class, I would prefer occasional in-class
    a. discussions or group problem-solving sessions.
    b. pauses that give opportunities to think or write about ideas presented in the lecture. ✓
22. On a multiple-choice test, I am more likely to
    a. run out of time.
    b. lose points because of not reading carefully or making careless errors. ✓
23. When I get directions to a new place, I prefer
    a. a map. ✓
    b. written instructions.
24. When I'm thinking about something I've read,
    a. I remember the incidents and try to put them together to figure out the themes.
    b. I just know what the themes are when I finish reading and then I have to back up and find the incidents that demonstrate them. ✓
25. When I get a new computer or VCR, I tend to
    a. plug it in and start punching buttons. ✓
    b. read the manual and follow instructions.
26. In reading for pleasure, I prefer
    a. something that teaches me new facts or tells me how to do something.
    b. something that gives me new ideas to think about. ✓
27. When I see a diagram or sketch in class, I am most likely to remember
    a. the picture. ✓
    b. what the instructor said about it.
28. It is more important to me that an instructor
    a. lay out the material in clear, sequential steps. ✓
    b. give me an overall picture and relate the material to other subjects.

SCORING SHEET: Use Table 3-1 to enter your scores.

1. Put 1s in the appropriate boxes in the table (e.g., if you answered **a** to Question 3, put a **1** in the column headed **a** next to the number **3**).
2. Total the 1s in the columns and write the totals in the indicated spaces at the base of the columns.

Table 3-1 Learning styles inventory scores.

Active/Reflective			Factual/Theoretical			Visual/Verbal			Linear/Holistic		
Q#	a	b	Q#	a	b	Q#	a	b	Q#	a	b
1		✓	2	✓		3	✓		4		✓
5	✓		6	✓		7	✓		8	✓	
9	✓		10	✓		11		✓	12		✓
13	✓		14		✓	15		✓	16		✓
17	✓		18		✓	19	✓		20	✓	
21		✓	22		✓	23		✓	24		✓
25	✓		26		✓	27	✓		28	✓	
Total	5	2	Total	4	3	Total	5	2	Total	5	2

3. For each of the four dimensions, circle your two scores on the bar scale and then fill in the bar between the scores. For example, if under "ACTV/REFL" you had 2 **a** and 5 **b** responses, you would fill in the bar between those two scores, as this sample shows:

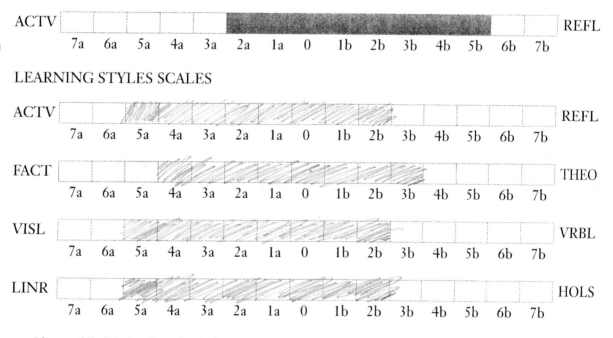

LEARNING STYLES SCALES

If your filled-in bar has the 0 close to its center, you are well balanced on the two dimensions of that scale. If your bar is drawn mainly to one side, you have a strong preference for that one dimension and may have difficulty learning in the other dimension.

Continue on to the next assessment. After you complete all three, the next section of the chapter will help you understand and make use of your results from each assessment.

Multiple Intelligences Theory

Howard Gardner, a Harvard University professor, has developed a theory called Multiple Intelligences. He believes there are at least eight distinct intelligences possessed by all people, and that every person has developed some intelligences more fully than others. Most people have experienced a time when they learned something very quickly and comfortably. Most have also had the opposite experience when, no matter how hard they tried, something they wanted to learn just would not sink in. According to the Multiple Intelligences Theory, when you find a task or subject easy, you are probably using a more fully developed intelligence; when you have more trouble, you may be using a less-developed intelligence.[4]

Following are brief descriptions of the focus of each of the intelligences. Study skills that reinforce each intelligence will be described later in the chapter.

- Verbal-Linguistic Intelligence—ability to communicate through language (listening, reading, writing, speaking)
- Logical-Mathematical Intelligence—ability to understand logical reasoning and problem solving (math, science, patterns, sequences)
- Bodily-Kinesthetic Intelligence—ability to use the physical body skillfully and to take in knowledge through bodily sensation (coordination, working with hands)
- Visual-Spatial Intelligence—ability to understand spatial relationships and to perceive and create images (visual art, graphic design, charts and maps)
- Interpersonal Intelligence—ability to relate to others, noticing their moods, motivations, and feelings (social activity, cooperative learning, teamwork)
- Intrapersonal Intelligence—ability to understand one's own behavior and feelings (independence, time spent alone)
- Musical Intelligence—ability to comprehend and create meaningful sound (music, sensitivity to sound)
- Naturalistic Intelligence—ability to understand features of the environment (interest in nature, environmental balance, ecosystem, stress relief brought by natural environments)

Please complete the following assessment of your multiple intelligences, called Pathways to Learning, developed by Joyce Bishop. It will help you determine which of your intelligences are most fully developed. Don't be concerned if some of your scores are low. That is true of most people, including your instructors and your authors!

Learning styles and multiple intelligences are guides to help you understand yourself. Instead of labeling yourself narrowly using one category or another, learn as much as you can about your preferences and how you can maximize your learning. Most people are a blend of styles and preferences, with one or two being dominant. In addition, you may change preferences depending on the situation. For example, a student might find it easy to take notes in outline style when the instructor lectures in an organized way.

TERMS

Intelligence
As defined by H. Gardner, an ability to solve problems or fashion products that are useful in a particular cultural setting or community.

TERMS

Kinesthetic
Coming from physical sensation cause by body movements and tensions.

PATHWAYS TO LEARNING

Directions: Rate each statement as follows: rarely 1; sometimes 2; usually 3; always 4.
Write the number of your response (1–4) in the box next to the statement and total each set of the six questions.

Developed by Joyce Bishop, Ph.D., and based upon Howard Gardner, *Frames of Mind: The Theory of Multiple Intelligences.*

[4] 1. I enjoy physical activities.
[1] 2. I am uncomfortable sitting still.
[3] 3. I prefer to learn through doing.
[2] 4. When sitting I move my legs or hands.
[4] 5. I enjoy working with my hands.
[1] 6. I like to pace when I'm thinking or studying.
[15] **TOTAL for Bodily/Kinesthetic**

[4] 7. I use maps easily.
[4] 8. I draw pictures/diagrams when explaining ideas.
[4] 9. I can assemble items easily from diagrams.
[2] 10. I enjoy drawing or photography.
[2] 11. I do not like to read long paragraphs.
[4] 12. I prefer a drawn map over written directions.
[20] **TOTAL for Visual/Spatial**

[1] 13. I enjoy telling stories.
[1] 14. I like to write.
[4] 15. I like to read.
[2] 16. I express myself clearly.
[2] 17. I am good at negotiating.
[1] 18. I like to discuss topics that interest me.
[11] **TOTAL for Verbal/Linguistic**

[3] 19. I like math in high school.
[2] 20. I like science.
[2] 21. I problem-solve well.
[1] 22. I question how things work.
[1] 23. I enjoy planning or designing something new.
[2] 24. I am able to fix things.
[12] **TOTAL for Logical/Mathematical**

[2] 25. I listen to music.
[2] 26. I move my fingers or feet when I hear music.
[1] 27. I have good rhythm.
[3] 28. I like to sing along with music.
[1] 29. People have said I have musical talent.
[1] 30. I like to express my ideas through music.
[11] **TOTAL for Musical**

[2] 31. I like doing a project with other people.
[1] 32. People come to me to help settle conflicts.
[3] 33. I like to spend time with friends.
[2] 34. I am good at understanding people.
[1] 35. I am good at making people feel comfortable.
[4] 36. I enjoy helping others.
[13] **TOTAL for Interpersonal**

[3] 37. I need quiet time to think.
[3] 38. I think about issues before I want to talk.
[4] 39. I am interested in self-improvement.
[4] 40. I understand my thoughts and feelings.
[4] 41. I know what I want out of life.
[2] 42. I prefer to work on projects alone.
[20] **TOTAL for Intrapersonal**

[4] 43. I enjoy nature whenever possible.
[1] 44. I think about having a career involving nature.
[3] 45. I enjoy studying plants, animals, or oceans.
[2] 46. I avoid being indoors except when I sleep.
[1] 47. As a child I played with bugs and leaves.
[3] 48. When I feel stressed, I want to be out in nature.
[14] **TOTAL for Naturalistic**

Write each of your eight intelligences in the column where it fits below. For each, choose the column that corresponds with your total in that intelligence.

\ Scores of 20–24 Highly Developed		Scores of 14–19 Moderately Developed		Scores Below 14 Underdeveloped	
Scores	Intelligences	Scores	Intelligences	Scores	Intelligences
20	visual/spatial	15	Bodily/kinesthetic	11	verbal/Linguistic
20	intrapersonal	14	naturalistic	12	logical/math
				11	musical
				13	interpersonal

Keys to Success, 2/e by Carter et al., 1998. Reprinted by permission of Prentice-Hall, Inc., Upper Saddle River, NJ.

However, if another instructor jumps from topic to topic, the student might choose to use the Cornell system or a think link (Chapter 7 goes into detail about note-taking styles).

The final assessment, through its evaluation of personality types, focuses on how you relate to others.

Personality Spectrum

A system that simplifies learning styles into four personality types has been developed by Joyce Bishop (1997). Her work is based on the Myers-Briggs and Keirsey and Bates theories discussed earlier in the chapter. The Personality Spectrum will give you a personality perspective on your learning styles. Please complete the following assessment.

When you have tallied your scores, plot them on Figure 3-1 to create a visual representation of your spectrum.

Your Personality Spectrum assessment can help you maximize your functioning at school and at work. Each personality type has its own abilities that improve work and school performance, suitable learning techniques, and ways of relating in interpersonal relationships. Table 3-2 explains what suits each type.

PERSONALITY SPECTRUM

Step 1. Rank all four responses to each question from **most** like you (4) to **least** like you (1). Place a 1, 2, 3, or 4 in each box next to the responses, and use each number only once per question.

1. I like instructors who
   - [4] a. tell me exactly what is expected of me.
   - [3] b. make learning active and exciting.
   - [2] c. maintain a safe and supportive classroom.
   - [1] d. challenge me to think at higher levels.

2. I learn best when the material is
   - [4] a. well organized.
   - [3] b. something I can do hands-on.
   - [2] c. about understanding and improving the human condition.
   - [1] d. intellectually challenging.

3. A high priority in my life is to
   - [4] a. keep my commitments.
   - [3] b. experience as much of life as possible.
   - [2] c. make a difference in other's lives.
   - [1] d. understand how things work.

4. Other people think of me as
   - [4] a. dependable and loyal.
   - [2] b. dynamic and creative.
   - [3] c. caring and honest.
   - [1] d. intelligent and inventive.

5. When I experience stress, I most likely
   - [1] a. do something to help me feel more in control.
   - [3] b. do something physical and daring.
   - [2] c. talk with a friend.
   - [4] d. go off by myself and think about my situation.

6. The greatest flaw someone can have is to be
   - [3] a. irresponsible.
   - [2] b. unwilling to try new things.
   - [4] c. selfish and unkind to others.
   - [1] d. an illogical thinker.

7. My vacations could best be described as
   - [2] a. traditional.
   - [1] b. adventuresome.
   - [3] c. pleasing to others.
   - [4] d. a new learning experience.

8. One word that best describes me is
   - [2] a. sensible.
   - [3] b. spontaneous.
   - [4] c. giving.
   - [1] d. analytical.

Step 2. Add up the total points for each column.

Total for (A)	Total for (B)	Total for (C)	Total for (D)
24	20	21	15
Organizer	Adventurer	Giver	Thinker

Step 3. Plot these numbers on the brain diagram on page 63.

From *Keys to Success: How to Achieve Your Goals*, 2/e by Carter et al., © 1998. Reprinted by permission of Prentice-Hall, Inc., Upper Saddle River, NJ.

Table 3-2

Personality spectrum at school and work.

PERSONALITY	STRENGTHS AT WORK AND SCHOOL	INTERPERSONAL RELATIONSHIPS
Organizer	◆ Can efficiently manage heavy work loads ◆ Good organizational skills ◆ Natural leadership qualities	◆ Loyal ◆ Dependable ◆ Traditional
Adventurer	◆ Adaptable to most changes ◆ Creative and skillful ◆ Dynamic and fast-paced	◆ Free ◆ Exciting ◆ Intense
Giver	◆ Always willing to help others ◆ Honest and sincere ◆ Good people skills	◆ Giving ◆ Romantic ◆ Warm
Thinker	◆ Good analytical skills ◆ Can develop complex designs ◆ Is thorough and exact	◆ Quiet ◆ Good problem solver ◆ Inventive

WHAT ARE THE BENEFITS OF KNOWING YOUR LEARNING STYLES?

Determining your learning styles profile takes work and self-exploration. For it to be worth your while, you need to understand what knowing your profile can do for you. The following sections will discuss benefits specific to study skills as well as more general benefits.

Study Benefits

Most students aim to maximize learning while minimizing frustration and time spent studying. If you know your particular learning style, you can use techniques that complement it. Such techniques take advantage of your highly developed areas while helping you through your less-developed ones. For example, say you perform better in smaller, discussion-based classes. When you have the opportunity, you might choose a course section that is smaller or that is taught by an instructor who prefers group discussion. You might also apply specific strategies to improve your retention in a lecture situation.

This section describes the techniques that tend to complement the strengths and shortcomings of each style. Students in Professor Soloman's program made many of these suggestions according to what worked for their own learning styles. Concepts from different assessments that benefit from similar strategies are grouped together. In Figure 3-2, you can see which styles tend to be dominant among students.

Remember that you may have characteristics from many different styles, even though some are dominant. Therefore, you may see suggestions for

Figure 3-1 — Personality spectrum—Thinking preferences & learning styles.

Place a dot on the appropriate number line for each of your four scores and connect the dots. A new shape will be formed inside each square. Color each shape in a different color.

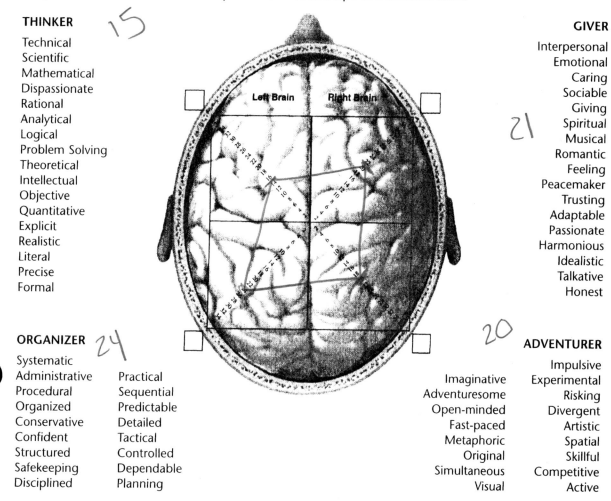

THINKER
Technical
Scientific
Mathematical
Dispassionate
Rational
Analytical
Logical
Problem Solving
Theoretical
Intellectual
Objective
Quantitative
Explicit
Realistic
Literal
Precise
Formal

GIVER
Interpersonal
Emotional
Caring
Sociable
Giving
Spiritual
Musical
Romantic
Feeling
Peacemaker
Trusting
Adaptable
Passionate
Harmonious
Idealistic
Talkative
Honest

ORGANIZER
Systematic
Administrative   Practical
Procedural   Sequential
Organized   Predictable
Conservative   Detailed
Confident   Tactical
Structured   Controlled
Safekeeping   Dependable
Disciplined   Planning

ADVENTURER
                 Impulsive
Imaginative   Experimental
Adventuresome   Risking
Open-minded   Divergent
Fast-paced   Artistic
Metaphoric   Spatial
Original   Skillful
Simultaneous   Competitive
Visual   Active

Source: *Understanding Psychology*, 3/e, by Morris, © 1996. Adapted by permission of Prentice-Hall, Inc., Upper Saddle River, NJ.

---

styles other than your dominant ones that may apply to you. What's important is that you use what works. Note the boxes next to the names of each style or type. In order to spot your best suggestions quickly, mark your most dominant styles or types by making check marks in the appropriate boxes.

Are You Active or Reflective?

Active learners ☐ include Bodily-Kinesthetic ☐ and Interpersonal ☐ learners as well as Adventurers. ☐ They like to apply the information to the real world, experience it in their own actions, or discuss or explain to others what they have learned.

Figure 3-2

Percentages of students with particular learning styles.

VISUAL	VERBAL
80%	20%

ACTIVE	REFLECTIVE
80%	20%

FACTUAL	THEORETICAL
70%	30%

LINEAR	HOLISTIC
85%	15%

Source: Barbara Soloman, North Carolina State University.

Student-suggested strategies for active learners:

- Study in a group in which members take turns explaining topics to each other and then discussing them.
- Think of practical uses for the course material.
- Pace and recite while you learn.
- Act out material or design games.
- Use flashcards with other people.
- Teach the material to someone else.

Reflective learners ☐ include Intrapersonal ☐ and Logical/Mathematical ☐ learners as well as Thinkers. ☐ They retain and understand information better after they have taken time to think about it.

Student-suggested strategies for reflective learners:

- Study in a quiet setting.
- When you are reading, stop periodically to think about what you have read.
- Don't just memorize material; think about why it is important and what it relates to, considering the causes and effects involved.
- Write short summaries of what the material means to you.

Are You Factual or Theoretical?

Factual learners ☐ and Organizers ☐ prefer concrete and specific facts, data, and detailed experimentation. They like to solve problems with standard methods and are patient with details. They don't respond well to surprises and unique complications that upset normal procedure. They are good at memorizing facts.

Student-suggested strategies for factual learners:

- Ask the instructor how ideas and concepts apply in practice.
- Ask for specific examples of the ideas and concepts.
- Brainstorm specific examples with classmates or by yourself.
- Think about how theories make specific connections with the real world.

Theoretical learners ☐ are often also logical/mathematical and prefer innovation and theories. They are good at grasping new concepts and big-picture ideas. They dislike repetition and fact-based learning. They are comfortable with symbols and abstractions, often connecting them with prior knowledge and experience. Most classes are aimed at theoretical learners.

Student-suggested strategies for theoretical learners:

- If a class deals primarily with factual information, try to think of concepts, interpretations, or theories that link the facts together.
- Because you become impatient with details, you may be prone to careless mistakes on tests. Read directions and entire questions before answering, and be sure to check your work.
- Look for systems and patterns that arrange facts in a way that makes sense to you.
- Spend time analyzing the material.

Are You Visual/Spatial or Verbal/Linguistic?

Visual/Spatial learners ☐ remember best what they see: diagrams, flowcharts, time lines, films, and demonstrations. They tend to forget spoken words and ideas. Classes generally don't include that much visual information. Note that although words written on paper or shown with an overhead projector are something you see, visual learners learn most easily from visual cues that don't involve words.

Student-suggested strategies for visual/spatial learners:

- Add diagrams to your notes whenever possible. Dates can be drawn on a time line; math functions can be graphed; percentages can be drawn in a pie chart.
- Organize your notes so that you can clearly see main points and supporting facts and how things are connected. You will learn more about different styles of note-taking in Chapter 7.
- Connect related facts in your notes by drawing arrows.
- Color-code your notes using different colored highlighters so that everything relating to a particular topic is the same color.

Verbal/Linguistic learners ☐ (often also interpersonal) remember much of what they hear and more of what they hear and then say. They benefit from discussion, prefer verbal explanation to visual demonstration, and learn effectively by explaining things to others. Because written words are processed as verbal information, verbal learners learn well through reading. The majority of classes, since they present material through the written word, lecture, or discussion, are geared to verbal learners.

Student-suggested strategies for verbal learners:

- Talk about what you learn. Work in study groups so that you have an opportunity to explain and discuss what you are learning.
- Read the textbook and highlight no more than 10 percent.
- Rewrite your notes.
- Outline chapters.
- Recite information or write scripts and debates.

Are You Linear or Holistic?

Linear learners ☐ **find it easiest to learn material presented in a logical, ordered progression.** They solve problems in a step-by-step manner. They can work with sections of material without yet fully understanding the whole picture. They tend to be stronger when looking at the parts of a whole rather than understanding the whole and then dividing it up into parts. They learn best when taking in material in a progression from easiest to more complex to most difficult. Many courses are taught in a linear fashion.

Student-suggested strategies for linear learners:

- If you have an instructor who jumps around from topic to topic, spend time outside of class with the instructor or a classmate who can help you fill the gaps in your notes.
- If class notes are random, rewrite the material according to whatever logic helps you understand it best.
- Outline the material.

Holistic learners ☐ **learn in fits and starts.** They may feel lost for days or weeks, unable to solve even the simplest problems or show the most basic understanding, until they suddenly "get it." They may feel discouraged when struggling with material that many other students seem to learn easily. Once they understand, though, they tend to see the big picture to an extent that others may not often achieve. They are often highly creative.

Student-suggested strategies for the holistic learner:

- Recognize that you are not slow or stupid. Don't lose faith in yourself. You will get it!
- Before reading a chapter, preview it by reading all the subheadings, summaries, and any margin glossary terms. The chapter may also start with an outline and overview of the entire chapter.
- Instead of spending a short time on every subject every night, try setting aside evenings for specific subjects and immerse yourself in just one subject at a time.
- Try taking difficult subjects in summer school when you are handling fewer courses.
- Try to relate subjects to other things you already know. Keep asking yourself how you could apply the material.

Study Techniques for Additional Multiple Intelligences

People who score high in the Musical/Rhythmic ☐ intelligence have strong memories for rhymes and can be energized by music. They often have a song running through their minds and find themselves tapping a foot or their fingers when they hear music.

Student-suggested strategies for musical/rhythmic people:

◆ Create rhymes out of vocabulary words.
◆ Beat out rhythms when studying.
◆ Play instrumental music while studying if it does not distract you, but first determine what type of music improves your concentration the most.
◆ Take study breaks and listen to music.
◆ Write a rap song about your topic.

Naturalistic learners ☐ feel energized when they are connected to nature. Their career choices and hobbies reflect their love of nature.

Student-suggested strategies for naturalistic people:

◆ Study outside whenever practical but only if it is not distracting.
◆ Explore subject areas that reflect your love for nature. Learning is much easier when you have a passion for it.
◆ Relate abstract information to something concrete in nature.
◆ Take breaks with something you love from nature—a walk, watching your fish, or a nature video. Use nature as a reward for getting other work done.

Study Techniques for Different Personality Types

The different personality types of the Personality Spectrum combine the learning styles and multiple intelligences you have explored. Table 3-3 shows learning techniques that benefit each type.

General Benefits

Although schools have traditionally favored verbal-linguistic students, there is no general advantage to one style over another. The only advantage is in discovering your profile through accurate and honest analysis. Following are three general benefits of knowing your learning styles:

1. You will have a better chance of avoiding problematic situations. If you don't explore what works best for you, you risk forcing yourself into career or personal situations that stifle your creativity, development, and happiness. Knowing how you learn and how you relate to the world can help you make smarter choices.
2. You will be more successful on the job. Your learning style is essentially your working style. If you know how you learn, you will be able to look

Table 3-3 Types and learning techniques.

PERSONALITY TYPES	RELATED LEARNING STYLES	LEARNING TECHNIQUES TO USE
Organizer	Factual, Linear	◆ Organize material before studying. ◆ Whenever possible, select instructors who have well-planned courses. ◆ Keep a daily planner and to-do list.
Adventurer	Active, Bodily-Kinesthetic	◆ Keep study sessions moving quickly. ◆ Make learning fun and exciting. ◆ Study with other Adventurers but also with Organizers.
Giver	Interpersonal	◆ Form study groups. ◆ Help someone else learn. ◆ Pick classes that relate to your interest in people.
Thinker	Reflective, Intrapersonal, Logical-Mathematical, Theoretical	◆ Study alone. ◆ Allow time to think about material. ◆ Pick classes and instructors who are intellectually challenging.

for an environment that suits you best and you'll be able to work effectively on work teams. This will prepare you for successful employment in the twenty-first century.

3. You will be better able to target areas that need improvement. The more you know about your learning styles, the more you will be able to pinpoint the areas that are more difficult for you. That has two advantages. One, you can begin to work on difficult areas, step by step. Two, when a task comes up requiring a skill that is tough for you, you can either take special care with it or suggest someone else whose style may be better suited to it.

Your learning style profile is one important part of self-knowledge. Next you will explore other important factors that help to define you.

How do you explore who you are?

You are an absolutely unique individual. Although you may share individual characteristics with others, your combination of traits is one-of-a-kind. It could take a lifetime to learn everything there is to know about yourself, because you are constantly changing. However, you can start by exploring these facets of yourself: self-perception, interests, habits, abilities, and limitations.

REAL WORLD PERSPECTIVE

How can I adjust my learning style to my instructors' teaching styles?

Patti Reed-Zweiger, South Puget Sound Community College, Tacoma, Washington

This last year I took a class in math that left me extremely stressed and exhausted. The way the teacher presented the material just didn't work for me. He threw out way too much information in a short period of time with little or no tools for completing the tasks. I really think he was unprepared. When he'd get to class, he'd fumble through his book for a while until he latched onto something to share. Sometimes, he'd spend the whole class answering a question or two about the previous homework and then, at the very last minute, give us a new assignment for the next class. We'd leave without any understanding of what we were to accomplish. It seems to me this teacher did very little teaching.

I'm a state trooper, so I'm used to handling enormous pressure, but in this case, nothing seemed to work. I'd leave in tears, class after class. This is frustrating for me. I'm forty years old and very confident, and yet in this class I felt like I was back in grade school again. I felt inadequate, foolish, and out of control. So much so that I would become sick to my stomach—nauseous. I wouldn't wish this experience on my worst enemy. What can I do to succeed in math and still maintain my self-esteem? At this point, I'm ready to drop it altogether.

Jacque Hall, University of Georgia, Terry College of Business

You're not alone. Math is frightening to most people. When I began taking math classes, I felt like a total failure. In fact, I dropped out of my Math 102 class. I just couldn't handle it. That's the first thing I'd recommend to you. Get out of the class if the teacher is not what you need. But make sure you talk with the teacher first and see if there's something the two of you can do to make the class successful for you. If you feel that it just won't work, let it go and try to find a better situation for yourself. Math is hard enough without subjecting yourself to inadequate teaching. I found that networking with other students on what classes and instructors to take really helped. The younger students always seem to know who the best teachers are.

If you can afford the additional time, I recommend you audit a class. If that isn't an option, hire a math tutor or take advantage of the math lab on a regular basis. Most importantly, remember that you are not a failure. And you're also not alone. I have felt a great deal of despair over math myself. I have seen people cry in class and others leave in total frustration. At some time or other, every student is going to run into a teacher or a classroom situation that leaves them feeling dissatisfied. Do your part by communicating with the teacher. If that doesn't work, move on. I'm glad I did.

Self-Perception

Having an accurate image of yourself is difficult. Unfortunately, many people do not have an accurate self-image. Feeling inadequate from time to time is normal, but a constantly negative self-perception can have destructive effects. Look at people you know who think that they are less intelligent, capable, or attractive than they really are. Observe how that shuts down their confidence and motivation. You do the same when you perceive yourself in a poor light.

Self-perception
How one views oneself, one's opinion of oneself.

For example, say you think you can't pass a certain course. Since you feel you don't have a chance, you don't put as much effort into the work for that course. Sure enough, at the end of the semester, you don't pass. The worst part is that you may see your failure as proof of your incapability, instead of realizing that you didn't allow yourself to try. This chain of events can occur in many situations. When it happens in the workplace, people lose jobs. When it happens in personal life, people lose relationships.

Refine your self-image so that it reflects more of your true self. These strategies might help.

- **Believe in yourself.** If you don't believe in yourself, others may have a harder time believing in you. Work to eliminate negative self-talk, for example, "I have a math block." Have faith in your abilities. When you set your goals, stick to them. Know that your mind and will are very powerful.
- **Talk to other people whom you trust.** People who know you well often have a more realistic perception of you than you do of yourself. Ask them what they think about your abilities and strengths.
- **Take personal time.** Stress makes having perspective on your life more difficult. Take time out to clear your mind and think realistically about who you are, who you want to be, and what is most important in your life.
- **Look at all of the evidence.** Mistakes can loom large in your mind. Consider what you do well and what you have accomplished as carefully as you consider your stumbles.

Building a positive self-perception is a lifelong challenge. If you maintain a bright but realistic vision of yourself, it will take you far along the road toward achieving your goals.

Interests

"The greatest discovery of any generation is that human beings can alter their lives by altering their attitudes of mind."

ALBERT SCHWEITZER

Taking some time now to explore your interests will help you later when you select a major and a career. You may be aware of many of your general interests already. For example, you can ask yourself:

- What areas of study do I like?
- What activities make me happy?
- What careers seem interesting to me?
- What kind of daily schedule do I like to keep (early riser, night owl)?
- What type of home and work environment do I prefer?

Interests play an important role in your life and in the workplace. Many people, however, do not take their interests seriously when choosing a career. Some make salary or stability their first priority. Some feel they have to take the first job that comes along. Some may not realize they can do better. Not considering what you are interested in may lead to an area of study or a job that leaves you unhappy, uninterested, or unfulfilled.

Choosing to consider your interests and happiness takes courage but brings benefits. Think about your life. You spend hours of time both attending classes and studying outside of class. You will spend at least eight hours a day, five or more days a week, up to fifty or more weeks a year as a working contributor to the world. Although your studies and work won't always make you deliriously happy, it is possible to spend your school and work time in a manner that suits you.

For instance, you may be a computer science major because everyone told you that you'd never get a job, or make money, as an artist—what you really wanted to study in college. Rather than choosing one or the other, combining them may be a possibility. You can continue as a computer science major and take plenty of art courses as electives. Plan on continuing your study of art as a lifelong pursuit *and* working as a computer scientist. Your various interests are not mutually exclusive; they can actually enhance each other. Creativity helps you in science, while science can help in your other pursuits.

Here are two positive effects of focusing on your interests.

1. You will have more energy. Think about how you feel when you are looking forward to seeing a special person, participating in a favorite sports activity, or enjoying some entertainment. When you're doing something you like, time seems to pass very quickly. You will be able to get much more done in a subject or career area that you enjoy.

2. You will perform better. When you were in high school, you probably got your best grades in your favorite classes and excelled in your favorite activities. That doesn't change as you get older. The more you like something, the harder you work at it—and the harder you work, the more you will improve.

Habits

A preference for a particular action that you do a certain way, and often on a regular basis or at certain times, is a habit. You might have a habit of showering in the morning, eating raisins, channel surfing with the TV remote control, hitting the snooze button on your clock, talking for hours on the phone, or studying late at night. Your habits reveal a lot about you. Some habits you consider to be good habits, and some may be bad habits.

Bad habits earn that title because they can prevent you from reaching important goals. Some bad habits, such as chronic lateness or smoking, cause obvious problems. Other habits, such as renting movies three times a week, may not seem bad until you realize that you needed to spend those hours studying. People maintain bad habits because they offer immediate, enjoyable rewards, even if later effects are negative. For example, going out to eat frequently may drain your budget, but at first it seems easier than shopping for food, cooking, and washing dishes.

Good habits are those that have positive effects on your life. You often have to wait longer and work harder to see a reward for good habits, which makes them harder to maintain. If you cut out fattening foods, you wouldn't lose weight in two days. If you reduced your nights out to gain study time, your grades wouldn't improve in a week. When you strive to maintain good habits, trust that the rewards are somewhere down the road.

Take time to evaluate your habits. Look at the positive and negative effects of each, and decide which are helpful and which harmful to you. Changing a habit can be a long process. Here are steps you can take to change a habit, to successfully make a behavior change.

1. **Be honest about your habits.** Admitting negative or destructive habits can be hard to do. You can't change a habit until you admit that it is a habit.
2. **Recognize the habit as troublesome.** Sometimes the trouble may not seem to come directly from the habit. For example, spending every weekend working on the house may seem important, but you may be overdoing it and ignoring friends and family members.
3. **Decide to change.** You might realize what your bad habits are but do not yet care about their effects on your life. Until you are convinced that you will receive something positive and useful from changing, your efforts will not get you far.
4. **Start today.** Don't put it off until after this week, after the family reunion, or after the semester. Each day lost is a day you haven't had the chance to benefit from a new lifestyle.
5. **Change one habit at a time.** Changing or breaking habits is difficult. Trying to spend more time with your family, reduce TV time, increase studying, and save more money all at once can bring on a fit of deprivation, sending you scurrying back to all your old habits. Easy does it.
6. **Reward yourself appropriately for positive steps taken.** If you earn a good grade, avoid slacking off on your studies the following week. Choose a reward that will not encourage you to stray from your target.
7. **Keep it up.** To have the best chance at changing a habit, be consistent for at least three weeks. Your brain needs time to become accustomed to the new habit. If you go back to the old habit during that time, you may feel like you're starting all over again.
8. **Don't get too discouraged.** Rarely does someone make the decision to change and do so without a setback or two. Being too hard on yourself might cause frustration that tempts you to give up and go back to the habit.

Abilities

Everyone's abilities include both strengths and limitations. And both can change. Examining both strengths and limitations is part of establishing the kind of clear vision of yourself that will help you to live up to your potential.

Strengths

As you think about your preferences, your particular strengths will come to mind, because you often like best the things you can do well. Some strengths seem to be natural—things you learned to do without ever having to work too hard. Others you struggled to develop and continue to work hard to maintain. Asking yourself these questions may help you define more clearly what your abilities are:

- What have I always been able to do well?
- What have others often praised about me?
- What do I like most about myself, and why?
- What is my learning style profile?
- What are my accomplishments—at home, at school, at work?

As with your preferences, knowing your abilities will help you find a job that makes the most of them. When your job requires you to do work you like, you are more likely to perform to the best of your ability. Keep that in mind as you explore career areas. Assessments and inventories that will help you further assess your abilities may be available at your school's career center or library. Once you know yourself, you will be more able to set appropriate goals.

Limitations

Being human means that nobody is perfect, and no one is good at everything. Everyone has limitations. However, that doesn't mean they are easy to take. Limitations can make you feel frustrated, stressed, or angry. You may feel as though no one else has the limitations you have, or that no one else has as many.

There are three ways to deal with your limitations. The first two—ignoring them or dwelling on them—are the most common. Neither is wise. The third way is to face them and to work to improve them while keeping the strongest focus on your abilities.

Ignoring your limitations can cause you to be unable to accomplish your goals. For example, say you are an active, global learner with a well-developed interpersonal intelligence. You have limitations in logical-mathematical intelligence and in linear thought. Ignoring that fact, you decide that you can make good money in computer programming, and you sign up for math and programming courses. You certainly won't fail automatically. However, if you ignore your limited ability in those courses and don't seek extra help, you may have more than a few stumbles.

Dwelling on your limitations can make you forget you have any strengths at all. This results in negative self-talk and a poor self-perception. Continuing the example, if you were to dwell on your limitations in math, you might very likely stop trying altogether.

Facing limitations and working to improve them is the logical response. A healthy understanding of your limitations can help you avoid troublesome situations. In the example, you could face your limitations in math and explore other career areas that use your more well-developed abilities and intelligences. If you decided to stick with computer technology, you could study an area of the field that focuses on management and interpersonal relationships. Or you could continue to aim for a career as a programmer, taking care to seek special help in areas that give you trouble.

Sabiduría

> "Discovery consists of seeing what everybody has seen and thinking what nobody has thought."
>
> ALBERT SZENT-GYÖRGYI

In Spanish, the term *sabiduría* represents the two sides of learning—both knowledge and wisdom. Knowledge—building what you know about how the world works—is the first part. Wisdom—deriving meaning and significance from knowledge, and deciding how to use that knowledge—is the second. As you continually learn and experience new things, the *sabiduría* you build will help you make knowledgeable and wise choices about how to lead your life.

Think of this concept as you discover more about how you learn and receive knowledge in all aspects of your life—in school, work, and personal situations. As you learn how your unique mind works and how to use it, you can more confidently assert yourself. As you expand your ability to use your mind in different ways, you can create lifelong advantages for yourself.

Chapter 3   Applications

Name _____   Date _____

KEY INTO YOUR LIFE
Opportunities to Apply What You Learn

 How Do You Learn Best?

Start by writing your scores next to each term.
Circle your highest preferences (largest numbers) for each assessment.

LEARNING STYLES INVENTORY	PATHWAYS TO LEARNING	PERSONALITY SPECTRUM
___ Active	___ Bodily-Kinesthetic	___ Organizer
___ Reflective	___ Visual-Spatial	___ Adventurer
___ Factual	___ Verbal-Linguistic	___ Giver
___ Theoretical	___ Logical-Mathematical	___ Thinker
___ Visual	___ Musical	
___ Verbal	___ Interpersonal	
___ Linear	___ Intrapersonal	
___ Holistic	___ Naturalist	

What positive experiences have you had at work and school that you can link to the strengths you circled?

_____
_____
_____
_____

What negative experiences have you had that may be related to your least-developed learning styles or intelligences?

_____
_____
_____
_____
_____

Making School More Enjoyable

List two required classes that you are not necessarily looking forward to taking. Discuss what parts of your learning style profile may relate to your lack of enthusiasm. Name learning styles–related study techniques that may help you get the most out of the class and enjoy it more.

CLASS	REASON FOR LACK OF ENTHUSIASM	LEARNING OR STUDY TECHNIQUES
1.		
2.		

Your Habits

You have the power to change your habits. List three habits that you want to change. Discuss the effects of each and how those effects keep you from reaching your goals.

HABIT	EFFECTS THAT PREVENT YOU FROM REACHING GOALS
1.	
2.	
3.	

Out of these three, put a star by the habit you want to change first. Write down a step you can take today toward overcoming that habit.

_____

_____

_____

What helpful habit do you want to develop in its place? For example, if your problem habit were a failure to express yourself when you are angry, a replacement habit might be to talk calmly about situations that upset you as soon as they arise. If you have a habit of cramming for tests at the last minute, you could replace it with a regular study schedule that allows you to cover your material bit by bit over a longer period of time.

_____

_____

_____

One way to help yourself abandon your old habit is to think about how your new habit will improve your life. List two benefits of your new habit.

1. _____
2. _____

Give yourself one month to complete your habit shift. Set a specific deadline. Keep track of your progress by indicating on a chart or calendar how well you did each day. If you avoided the old habit, write an X below the day. If you used the new one, write an N. Therefore, a day when you only avoided the old habit will have an X; a day when you did both will have both letters; and a day when you did neither will be left blank. You can use the chart below or mark your own calendar. Try pairing up with another student and arranging to check up on each other's progress.

1	2	3	4	5	6	7	8	9	10	11	12	13	14	15	16
17	18	19	20	21	22	23	24	25	26	27	28	29	30	31	

Don't forget to reward yourself for your hard work. Write here what your reward will be when you feel you are on the road to a new and beneficial habit.

_____
_____
_____

KEY TO SELF-EXPRESSION
Discovery Through Journal Writing

To record your thoughts, use a separate journal or the lined pages at the end of the chapter.

Your Learning Style Profile Discuss the insights you have gained, through exploring your learning style profile, about your strengths and struggles at school and work. What new strengths have come to your attention? What struggles have you become aware of that you couldn't explain before? Talk about how your insights may have changed the way you see yourself as a science major.

Name _____ Date _____

Journal

Journal

Name _____ Date _____

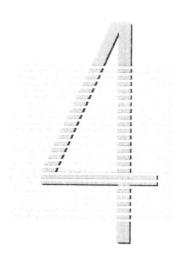

Goal Setting and Time Management

Mapping Your Course

People dream of what they want out of life, but not everyone knows how to turn dreams into reality. Often dreams seem far off in time, too difficult, or even completely unreachable. You can build paths to your dreams, however, by identifying the goals you need to achieve, one by one, to arrive at your destination. When you set goals, prioritize, and manage your time effectively, you increase your ability to take those steps to achieve your long-term goals.

This chapter explains how taking specific steps toward your goals in science can help you turn your dreams into reality. You will explore how your values relate to your goals, how to create a framework for your life's goals, how to set long-term and short-term goals, and how to set priorities. The section on time management will discuss how to translate those goals into daily, weekly, monthly, and yearly steps. Finally, you will explore the effects of procrastination.

In this chapter, you will explore answers to the following questions:

- What defines your values?
- How do you set and achieve goals?
- What are your priorities?
- How can you manage your time?
- Why is procrastination a problem?

WHAT DEFINES YOUR VALUES?

TERMS

Values Principles or qualities that one considers important, right, or good.

Your personal **values** are the beliefs that guide your choices. Examples of values include family togetherness, a good education, caring for others, working to protect the environment, and worthwhile employment. The sum total of all your values is your *value system.* You demonstrate your particular value system in the priorities you set, how you communicate with others, your family life, your educational and career choices, even the material things with which you surround yourself, and your lifestyle.

Choosing and Evaluating Values

Examining the sources of your values can help you define those values, trace their origin, and question the reasons why you have adopted them. Value sources, however, aren't as important as the process of considering each value carefully to see if it makes sense to you. Some of your current values may have come from television or other media but still ring true. Some may come from what others have taught you. Some you may have constructed from your own personal experience and opinion. You make the final decision about what to value, regardless of the source.

Each individual value system is unique, even if many values come from other sources. Your value system is yours alone. Your responsibility is to make sure that your values are your own choice, not the choice of others. Make value choices for yourself based on what feels right for you, for your life, and for those who are touched by your life.

"The important thing in science is not so much to obtain new facts as to discover new ways of thinking about them."
WILLIAM LAWRENCE BRAGG

You can be more sure of making choices that are right for you if you try to always question and evaluate your values. Before you adopt a value, ask yourself: Does it feel right? What effects might it have on my life? Am I choosing it to please someone else, or is it truly my choice? Values are a design for life, and you are the one who has to live the life you design.

Because life changes and new experiences may bring a change in values, try to continue to evaluate values as time goes by. Periodically evaluate the effects that having each value has on your life, and see if a shift in values might suit your changing circumstances. For example, losing your sight may cause you to value your hearing intensely. The difficulty of a divorce may have a positive result: a new value of independence and individuality. After growing up in a homogeneous community, a student who meets other students from unfamiliar backgrounds may learn a new value of living in a diverse community. Your values will grow and develop as you do if you continue to think them through.

How Values Relate to Goals

Understanding your values will help you set your science career and personal goals, because the most ideal goals help you achieve what you value. Values of financial independence or simple living may generate goals, such as working while going to school and keeping credit card debt low, that reflect the value. If you value helping others, try to make time for volunteer work.

Goals enable you to put values into practice. When you set and pursue goals that are based on values, you demonstrate and reinforce values through taking action. The strength of those values, in turn, reinforces your goals. You will experience a much stronger drive to achieve if you build goals around what is most important to you.

How do you set and achieve goals?

A goal can be something as concrete as taking biochemistry next semester or as abstract as working to control your temper. When you set goals and work to achieve them, you engage your intelligence, abilities, time, and energy in order to move ahead. From major life decisions to the tiniest day-to-day activities, setting goals will help you define how you want to live and what you want to achieve.

Goal
An end toward which effort is directed; an aim or intention.

Paul Timm, a best-selling author and teacher who is an expert in self-management, feels that focus is a key ingredient in setting and achieving goals. "Focus adds power to our actions. If somebody threw a bucket of water on you, you'd get wet, and probably get mad. But if water was shot at you through a high-pressure nozzle, you might get injured. The only difference is focus."[1] Each part of this section will explain ways to focus your energy through goal setting. You can set and achieve goals by defining a personal mission statement, placing your goals in long-term and short-term time frames, evaluating goals in terms of your values, and linking your goals to your values.

Identifying Your "Personal Mission Statement"

Some people go through their lives without ever really thinking about what they can do or what they want to achieve. If you choose not to set goals or explore what you want out of life, you may look back on your past with a sense of emptiness. You may not know what you've done or why you did it. By periodically taking a few steps back and thinking about where you've been and where you want to be, you'll live more consciously.

One helpful way to determine your general direction is to write a *personal mission statement*. Dr. Stephen Covey, author of the best-seller *The Seven Habits of Highly Effective People*, defines a mission statement as a philosophy that outlines what you want to be (character), what you want to do (contributions and achievements), and the principles by which you live. Dr. Covey compares the personal mission statement to the Constitution of the United States, a statement of principles that gives this country guidance and standards in the face of constant change.[2]

Your personal mission isn't written in stone. It should change as you move from one phase of life to the next—from single person to spouse, from parent to single parent to caregiver of an older parent. Stay flexible and reevaluate your personal mission from time to time.

Here is an example of author Janet Katz's personal mission statement:

> My mission is to uphold the nursing profession's value of advocating for those in need by promoting the health and well-being of people of all ages, backgrounds, and economic levels through local and international community health efforts. I intend to celebrate life through service to others and caring for myself and my family.

Here is a mission statement from Immunex Corporation, a biotechnology company based in Seattle, Washington:

> Immunex is a biopharmaceutical company dedicated to developing immune system science to protect human health. The company's products offer hope to patients with cancer, inflammatory and infectious diseases.

Another example is from The Nature Conservancy, a nonprofit organization responsible for the protection of more than 10 million acres in the United States and Canada and partnerships in Latin America, the Caribbean, the Pacific, and Asia:

> The mission of The Nature Conservancy is to preserve plants, animals, and natural communities that represent the diversity of life on Earth by protecting the lands and waters they need to survive.

You will have an opportunity to write your own personal mission statement at the end of this chapter. Writing a mission statement is much more than an in-school exercise. It is truly for you. Thinking through your personal mission can help you begin to take charge of your life. It helps to put you in control instead of allowing circumstances and events to control you. If you frame your mission statement carefully so that it truly reflects your goals, it can be your guide in everything you do.

Placing Goals in Time

Everyone has the same 24 hours in a day, but it often doesn't feel like enough. Have you ever had a busy day flash by so quickly that it seems you accomplished nothing? Have you ever felt that way about a longer period of time, like a month or even a year? Your commitments can overwhelm you unless you decide how to use time to plan your steps toward goal achievement.

If developing a personal mission statement establishes the big picture, placing your goals within particular time frames allows you to bring individual areas of that picture into the foreground. Planning your progress step-by-step will help you maintain your efforts over the extended time period often needed to accomplish a goal. Goals fall into two categories: long-term and short-term.

Setting Long-Term Goals

Establish first the goals that have the largest scope; that is, the *long-term goals* that you aim to attain over a lengthy period of time, up to a few years or more. As a student, you know what long-term goals are all about. You have set yourself a goal to attend school and earn a degree or certificate. Becoming educated is an admirable goal that takes a good number of years to reach.

Some long-term goals are lifelong, such as a goal to continually learn more about yourself and the world around you. Others have a more definite end, such as a goal to complete a course successfully. To determine your long-term goals, think about what you want out of your professional, educational, and personal life. Here is Janet Katz's long-term goal statement.

Janet's Goals: To accomplish my mission through writing books and journal articles, teaching nursing students, and developing improved methods for promoting the profession of nursing and its values. To create and maintain a lifestyle that is conducive to my own physical and mental health and that of my family.

For example, you may establish long-term goals such as these:

- I will graduate from college with the degree I most desire, having learned and understood as much as I could in a wide range of subjects.
- I will build my science inquiry and research skills through work, volunteering, and internships or through relationships with course instructors, other professionals in my field of interest, classmates, and co-workers.

Long-term goals can change later in your life. To begin long-term goal setting, start with next year. Deciding what you want to accomplish in the next year and writing it down will help you to focus clearly on productive actions. These goals are not like New Year's resolutions; they are based on what you really are willing to work toward and accomplish. Janet's goals focused on what she wanted to accomplish next year.

1. Finish current book project and begin investigating two other book ideas by talking to editors.
2. To exercise daily, eat six to seven servings of fruits and vegetables per day, and read books for my own enjoyment. Make time to reflect on my life and the life around me.

In the same way that Janet's goals are tailored to her personality and interests, your goals should reflect who you are. Personal missions and goals are as unique as each individual. Continuing the example above, you might adopt these goals for the coming year:

- I will earn passing grades in all my classes.
- I will volunteer and assist my engineering professor in her current research project.

Setting Short-Term Goals

When you divide your long-term goals into smaller, manageable goals that you hope to accomplish within a relatively short time, you are setting *short-term goals*. Short-term goals narrow your focus, helping you to maintain your progress toward your long-term goals. They are the steps that take you where you want to go. Say you have set the two long-term goals you just read in the previous section. To stay on track toward those goals, you may want to accomplish these short-term goals in the next six months:

- I will pass Chemistry I, so that I can move on to Chemistry II.
- I will read three journal articles pertinent to my engineering professor's research project.

These same goals can be broken down into even smaller parts, such as in one month:

- I will complete the last week's lab write-up and do the reading for the next week's lab by Sunday night of each week.
- I will read a research article from *Journal of Materials in Civil Engineering* and prepare a brief report on it for next month's Engineering Club's brown bag seminar.

In addition to monthly goals, you may have short-term goals that extend a week, a day, or even a couple of hours in a given day. Take as an example the article you planned to present for next month's Engineering Club's brown bag seminar. Such short-term goals may include the following:

- Three weeks from now: Attend the seminar ready to present a 10-minute clear summary of a research article from *Journal of Materials in Civil Engineering*.
- Two weeks from now: Have a final draft of the presentation and ask another Engineering Club member to review it.
- One week from now: Have a first draft of an outline ready, and ask the seminar instructor to read it.
- Today by the end of the day: Find an interesting research article, and submit it to the seminar instructor.
- By 3 P.M. today: Brainstorm ideas of topics, and go to the library to start searching *Journal of Materials in Civil Engineering*.

As you consider your long-term and short-term goals, notice how all of your goals are linked to one another. As Figure 4-1 shows, your long-term goals establish a context for the short-term goals. In turn, your short-term goals make the long-term goals seem clearer and more reachable. The whole system works to keep you on track.

Figure 4-1   Linking goals together.

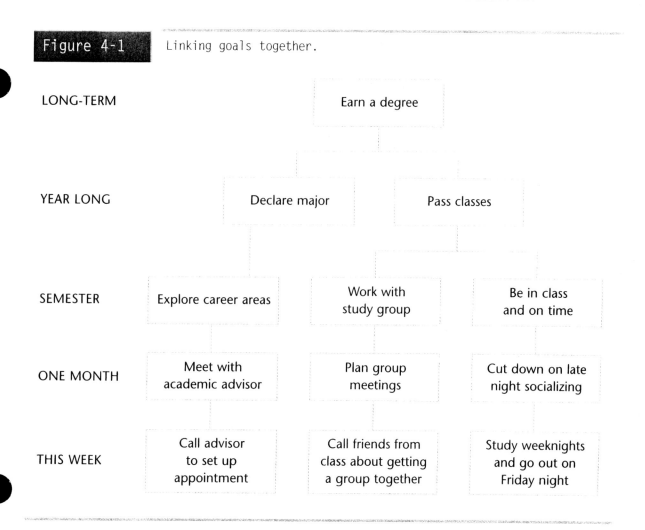

Linking Goals With Values

If you are not sure how to start formulating your mission, look to your values to guide you. Define your mission and goals based on what is important to you.

If you value physical fitness, your mission statement might emphasize your commitment to staying in shape throughout your life. Your long-term goal might be to run a marathon, while your short-term goals might involve your weekly exercise and eating plans. Similarly, if you value a close family, your personal mission might emphasize how you want to maintain family ties and stability. In this case, your long-term goals might involve finding a job that allows for family time or living in a town close to your parents. Your short-term goals may focus on helping your son learn a musical instrument or having dinner with your family at least twice a week.

"Even if you're on the right tract, you'll get run over if you just sit there."

WILL ROGERS

Current and Personal Values Mean Appropriate Goals

When you use your values as a compass for your goals, make sure the compass is pointed in the direction of your real feelings. Watch out for the following two pitfalls that can occur.

Setting goals according to other peoples' values. Friends or family may encourage you to strive for what they think you should value, rather than what is right for you. If you follow their advice without believing in it, you may have a harder time sticking to your path. For example, someone who attends school primarily because a parent or spouse thought it was right may have less motivation and initiative than someone who made an independent decision to become a student. Look hard at what you really want, and why. Staying in tune with your own values will help you make decisions that are right for you.

Setting goals that reflect values you held in the past. What you felt yesterday may no longer apply, because life changes can alter your values. The best goals reflect what you believe today. For example, a person who has been through a near-fatal car accident may experience a dramatic increase in how he or she values time with friends and family, and a drop in how he or she values material possessions. Someone who survives a serious illness may value healthy living above all else. Keep in touch with your life's changes so your goals can reflect who you are.

Values Can Help You Identify Educational Goals

Education is a major part of your life right now. In order to define a context for your school goals, explore what you value about pursuing an education. People have many reasons for attending college. You may identify with one or more of the following possible reasons:

- I want to earn a higher salary.
- I want to build marketable skills.
- My supervisor at work says that a degree will help me move ahead in my career.
- Most of my friends were going.
- I want to be a student and learn all that I can.
- It seems like the only option for me right now.
- I am recently divorced and need to find a way to earn money.
- Everybody in my family goes to college; it's expected.
- I don't feel ready to jump into the working world yet.
- I got a scholarship.
- My friend loves her job and encouraged me to take courses in the field.
- My parent (or a spouse or partner) pushed me to go to college.
- I am pregnant and need to increase my skills so I can provide for my baby.
- I am studying for a specific career.
- I don't really know.

All of these answers are legitimate, even the last one. Being honest with yourself is crucial if you want to discover who you are and what life paths make sense for you. Whatever your reasons are for being in school, you are at the gateway to a journey of discovery.

It isn't easy to enroll in college, pay tuition, decide what to study, sign up for classes, gather the necessary materials, and actually get yourself to the school and into the classroom. Many people drop out at different places along the way, but somehow your reasons have been compelling enough for you to have arrived at this point. Thinking about why you value your education will help you stick with it.

Achieving goals becomes easier when you are realistic about what is possible. Setting priorities will help you make that distinction.

WHAT ARE YOUR PRIORITIES?

When you set a priority, you identify what's important at any given moment. *Prioritizing* helps you focus on your most important goals, even when they are difficult to achieve. If you were to pursue your goals in no particular order, you might tackle the easy ones first and leave the tough ones for later. The risk is that you might never reach for goals that are important to your success. Setting priorities helps you focus your plans on accomplishing your most important goals.

TERMS

Priority
An action or intention that takes precedence in time, attention, or position.

To explore your priorities, think about your personal mission and look at your goals in the five life areas: personal, family, school/career, finances, and lifestyle. These five areas may not all be equally important to you right now. At this stage in your life, which two or three are most critical? Is one particular category more important than others? How would you prioritize your goals from most important to least important?

You are a unique individual, and your priorities are yours alone. What may be top priority to someone else may not mean that much to you, and vice versa. You can see this in Figure 4-2, which compares the priorities of two very different students. Each student's priorities are listed in order, with the first priority at the top and the lowest priority at the bottom.

First and foremost, your priorities should reflect your personal goals. In addition, they should reflect your relationships with others. For example, if you are a parent, your children's needs will probably be high on the priority list. You may decide to go back to school so you can get a better job, earn more money, and give them a better life. If you are in a committed relationship, you may consider the needs of your partner. You may schedule your classes so that you and your partner are home together as often as possible. Even as you consider the needs of others, though, never lose sight of your personal goals. Be true to your goals and priorities so that you can make the most of who you are.

Setting priorities moves you closer to accomplishing specific goals. It also helps you begin planning to achieve your goals within specific time frames. Being able to achieve your goals is directly linked to effective time management.

HOW CAN YOU MANAGE YOUR TIME?

Time is one of your most valuable and precious resources. Unlike money, or opportunity, or connections, time doesn't discriminate—everyone has the same twenty-four hours in a day, every day. Your responsibility and your

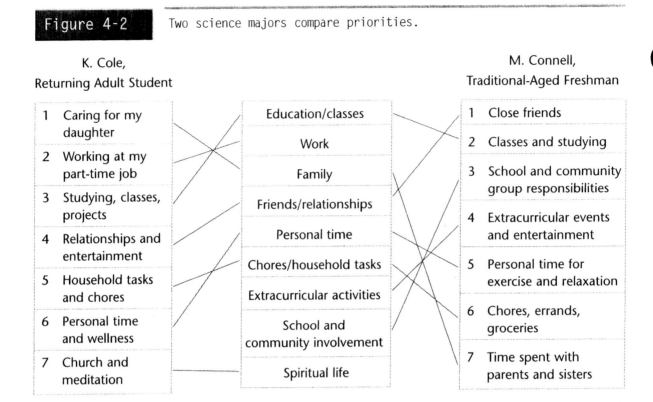

Figure 4-2 Two science majors compare priorities.

potential for success lie in how you use yours. You cannot manipulate or change how time passes, but you can spend it taking steps to achieve your goals. Efficient time management helps you achieve your goals in a steady, step-by-step process.

Science classes can be time-consuming because they often include lab time and many hours of study. As a science major, you can promote your own success by learning to creatively and accurately manage your time.

People have a variety of different approaches to time management. Your learning style (explained in more detail in Chapter 3) can help you identify the particular way you currently use your time. For example, factual and linear learners tend to organize activities within a framework of time. Because they stay aware of how long it takes them to do something or travel somewhere, they are usually prompt. Theoretical and holistic learners tend to miss the passing of time while they are busy thinking of something else. Because they focus on the big picture, they may neglect details such as structuring their activities within available time. They frequently lose track of time and can often be late without meaning to be.

Time management, like physical fitness, is a lifelong pursuit. No one can plan a perfect schedule or build a terrific physique and then be "done." You'll work at time management throughout your life, and it can be tiring. Your ability to manage your time will vary with your mood, your stress level, how busy you are, and other factors. You're human; don't expect perfection. Just do your best. Time management involves building a schedule and making your schedule work through lists and other strategies.

REAL WORLD PERSPECTIVE

Carolyn Christina Moos, Stanford University, California, Human Biology Major

I entered Stanford as a freshman without any idea science would become such a large part of my academic pursuit. I started out interested in psychology and communications, but I soon found that science was at the core of many disciplines due to its systematic study and analysis. I was interested in being a sports physiologist, including nutrition and sports psychology, and now as a sophomore, I've found my niche in the Human Biology program.

Time management is integral to success. As an athlete (women's basketball team), I am forced to use my time wisely or I will not achieve my goals. This means that after practice I may come home to friends and roommates that feel like taking a break or relaxing. Then I have to tell myself that I have to start working even though I am tired. The key is to do this to stay on top and not fall behind. I thoroughly enjoy human biology, which makes it much more enjoyable to study. It is not an easy route, but if you're pursuing something you love, the time is definitely worth it.

So, how do you do this? The key is an eight letter word: WEEKENDS. For many, weekends are the time to kick back, but honestly, it is the time to get ahead. By studying on weekends, you can minimize stress in the upcoming week. Since my schedule includes games at home and away, I have to make a conscious effort to stay ahead. Everyone has to maintain a good GPA to stay on the team.

Playing basketball and pursuing a time-demanding subject like science work well together because playing basketball requires a lot of discipline. You learn to be punctual; you know what it means to work hard; and you know what it means to stretch yourself beyond your perceived limits. The study of science also helps me understand my sport. I can understand what my body is doing as I run up and down, cut left to right, jump, and sweat. This also allows me to connect with my body better. Basketball is also a good balance with academic demands because the physical exertion helps me to focus when I study.

My key advice to a beginning science student is that you consider seeking something you are passionate about, realizing that if you open the door, opportunity will flourish. Don't let yourself be held back by issues like time, intensity, or intimidation. The only things that are really intimidating are the ones you don't know about. I came to Stanford with a narrow mind, but when I opened my eyes, I realized that science played a large role in rounding out my knowledge in sports. I know science will help me in my future endeavors, and I encourage everyone who takes science courses to prepare for a positive academic experience.

Building a Schedule

Being in control of how you manage your time is a key factor in taking responsibility for yourself and your choices. When you plan your activities with an eye toward achieving your most important goals, you are taking personal responsibility for how you live. Building a schedule helps you be responsible.

Just as a road map helps you travel from place to place, a *schedule* is a time-and-activity map that helps you get from the beginning of the day (or week, or month) to the end as smoothly as possible. A written schedule helps you gain control of your life. Schedules have two major advantages: They allocate segments of time for the fulfillment of your daily, weekly, monthly, and longer-term goals, and they serve as a concrete reminder of tasks, events, due dates, responsibilities, and deadlines. Few moments are more stressful than suddenly realizing you have forgotten to pick up a prescription, take a test, or be on duty at work. Scheduling can help you avoid events like these.

Keep a Date Book

Gather the tools of the trade: a pen or pencil and a *date book* (sometimes called a planner). Some of you already have date books and may have used them for years. Others may have had no luck with them or have never tried. Even if you don't feel you are the type of person who would use one, give it a try. A date book is indispensable for keeping track of your time. Paul Timm says, "Most time management experts agree that rule number one in a thoughtful planning process is: Use some form of a planner where you can write things down."

There are two major types of date books. The *day-at-a-glance* version devotes a page to each day. While it gives you ample space to write the day's activities, this version makes it difficult to see what's ahead. The *week-at-a-glance* book gives you a view of the week's plans but has less room to write per day. If you write out your daily plans in detail, you might like the day-at-a-glance version. If you prefer to remind yourself of plans ahead of time, try the book that shows a week's schedule all at once. Some date books contain additional sections that allow you to note plans and goals for the year as a whole and for each month. You can also create your own sheets for yearly and monthly notations in a notepad section, if your book has one, or on plain paper that you can then insert into the book.

Another option to consider is an *electronic planner*. These are compact minicomputers that can hold a large amount of information. You can use them to schedule your days and weeks, make to-do lists, and create and store an address book. Electronic planners are powerful, convenient, and often fun. On the other hand, they certainly cost more than the paper version, and you can lose a lot of important data if something goes wrong with the computer inside. Evaluate your options and decide what you like best.

Set Weekly and Daily Goals

The most ideal time management starts with the smallest tasks and builds to bigger ones. Setting short-term goals that tie in to your long-term goals lends the following benefits:

- Increased meaning for your daily activities
- Shaping your path toward the achievement of your long-term goals
- A sense of order and progress

For college students as well as working people, the week is often the easiest unit of time to consider at one shot. Weekly goal setting and planning allows you to keep track of day-to-day activities while giving you the larger perspective of what is coming up during the week. Take some time before each week starts to remind yourself of your long-term goals. Keeping long-term goals in mind will help you determine related short-term goals you can accomplish during the week to come.

Figure 4-3 shows parts of a daily schedule and a weekly schedule.

Link Daily and Weekly Goals With Long-Term Goals

After you evaluate what you need to accomplish in the coming year, semester, month, week, and day in order to reach your long-term goals, use your schedule to record those steps. Write down the short-term goals that will enable

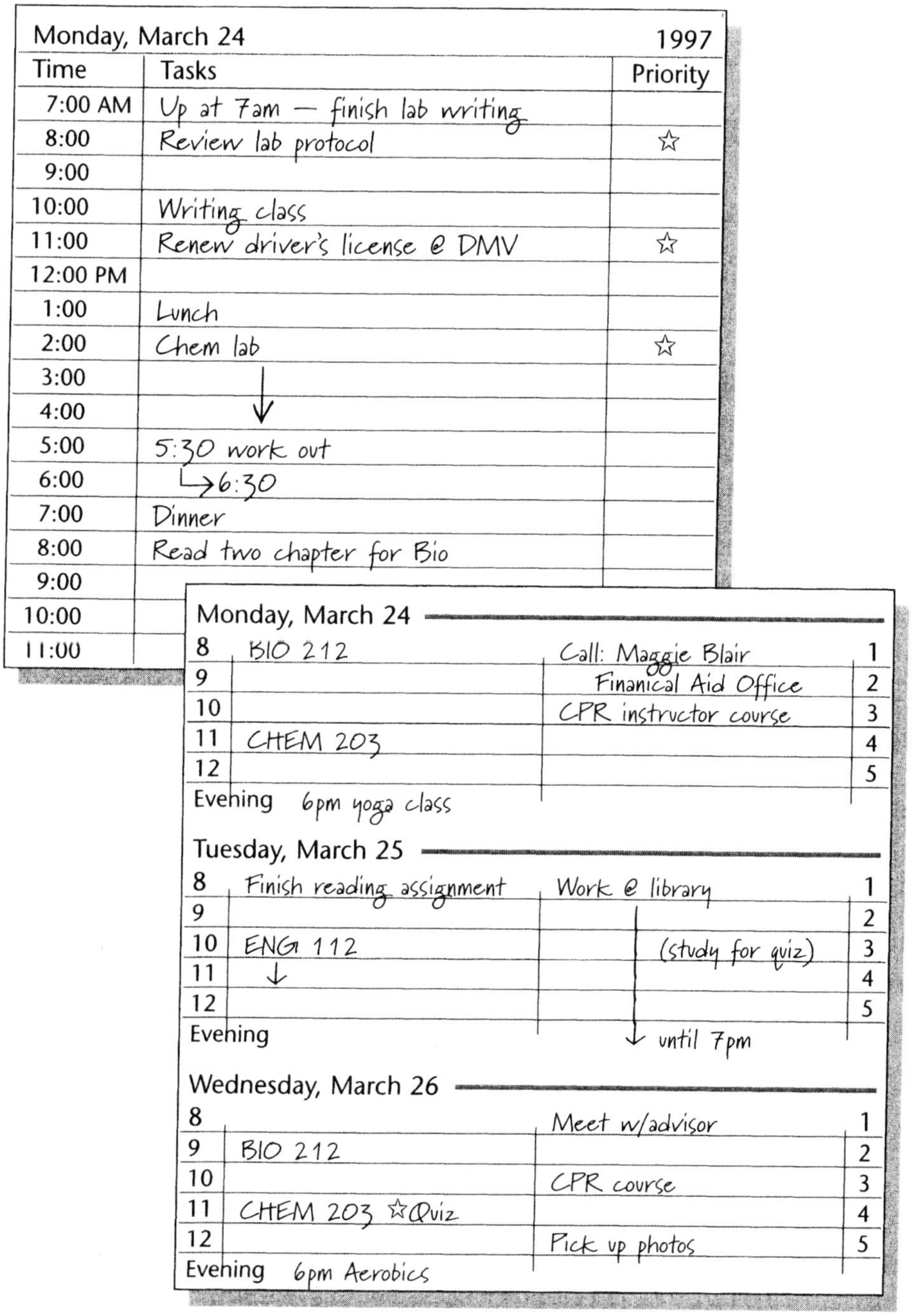

Figure 4-3

Daily and weekly schedules.

Monday, March 24		1997
Time	Tasks	Priority
7:00 AM	Up at 7am — finish lab writing	
8:00	Review lab protocol	☆
9:00		
10:00	Writing class	
11:00	Renew driver's license @ DMV	☆
12:00 PM		
1:00	Lunch	
2:00	Chem lab	☆
3:00	↓	
4:00		
5:00	5:30 work out	
6:00	↳ 6:30	
7:00	Dinner	
8:00	Read two chapter for Bio	
9:00		
10:00		
11:00		

Monday, March 24			
8	BIO 212	Call: Maggie Blair	1
9		Finanical Aid Office	2
10		CPR instructor course	3
11	CHEM 203		4
12			5
Evening	6pm yoga class		

Tuesday, March 25			
8	Finish reading assignment	Work @ library	1
9			2
10	ENG 112	(study for quiz)	3
11	↓		4
12			5
Evening		↓ until 7pm	

Wednesday, March 26			
8		Meet w/advisor	1
9	BIO 212		2
10		CPR course	3
11	CHEM 203 ☆Quiz		4
12		Pick up photos	5
Evening	6pm Aerobics		

you to stay on track. Here is how a student might map out two different goals over a year's time.

This year:	Complete enough courses to graduate.
	Improve my physical fitness.
This semester:	Complete my applied mathematics course with a B average or higher.
	Lose 10 pounds and exercise regularly.
This month:	Set up study group schedule to coincide with quizzes.
	Begin walking and weight lifting.
This week:	Meet with study group; go over material for Friday's quiz.
	Go for a fitness walk three times; go to weight room twice.
Today:	Go over Chapter 3 in applied mathematics and do problems 1–5.
	Walk for 40 minutes.

Prioritize Goals

Prioritizing enables you to use your date book with maximum efficiency. On any given day, the necessity of completing your goals will vary. Record your goals first, and then label them according to how important they are to complete using these categories: priority 1, priority 2, and priority 3. Identify these categories using any code that makes sense to you. Some people use numbers, as above. Some use letters (A, B, C). Some write activities in different colors according to priority level. Some use symbols (*, +, −).

Priority 1 activities are the most necessary or critical to complete. They may include attending class, picking up a child from day care, putting gas in the car, and paying bills.

Priority 2 activities are part of your routine. Examples include grocery shopping, working out, participating in a school organization, or cleaning. Priority 2 tasks are important but more flexible than priority 1 activities.

Priority 3 activities are those you would like to do but can reschedule without much sacrifice. Examples might be a trip to the mall, a visit to a friend, a social phone call, a sports event, a movie, or a hair appointment. As much as you would like to accomplish them, you don't consider them urgent. Many people don't enter priority 3 tasks in their date books until they are sure they have time to get them done. You may want to list priority 3 tasks separately and refer to the list when you have some extra time.

Prioritizing your activities is essential for two reasons. First, some activities are more necessary to complete than others, and effective time management requires that you focus most of your energy on priority 1 items. Second, looking at all your priorities helps you plan when you can get things done. Often, it's not possible to get all your priority 1 activities done early in the day, especially if these activities involve scheduled classes or meetings. Prioritizing helps you set priority 1 items and then schedule priority 2 and 3 items around them as they fit.

> "The right time is any time that one is still so lucky as to have. . . . Live!"
>
> HENRY JAMES

REAL WORLD PERSPECTIVE

How can I stay focused on my goals?

Karin Lounsbury, Gonzaga University, Spokane, Washington

I decided to return to school when I had just turned forty. I didn't like feeling dependent on my husband for my financial security so I thought that I'd do something about it. I also did it for my two children. My marriage had been shaky for quite a few years and I was scared to death that I wouldn't be able to provide for them on my own. Even though I'd worked in the business world for a long time, the salary was never very good. I was over-experienced and underpaid. I thought that by completing my education, I could find a great job that allowed me to support my family. Although I knew that college would be challenging, I wasn't concerned with the work load—I'm used to carrying a lot of responsibilities. In fact, probably more than most people. Besides my two young children, I'm married to a man who lost both his legs in the Vietnam war. He's in a wheelchair, which means a lot of extra work falls on my shoulders.

This last few months everything seems to be falling apart. My husband and I decided to get a divorce; my son has been struggling at school; my mother was just diagnosed with cancer; and I feel like I can hardly keep my head above water. All of this is taking a toll on my grades. I'm usually so emotionally and physically exhausted by the end of the day, I just don't have the energy to put into my work. When I'm at school, I'm distracted thinking about the future. I don't want to drop out of school but I also don't want my kids to suffer when they need me so badly. How can I get through this difficult time and still accomplish my educational goals?

Shirley Williamson, University of Georgia

To begin with, I want to encourage you to hold onto your dream of finishing your education. Even though there are probably going to be some very cloudy days ahead, don't give up. The long-term rewards are worth all the extra effort it's going to take for a while. If you could lighten your academic load in any way, I think it would be wise to do so. It might take you a little longer to graduate, but you and your children will appreciate the extra time you get with one another. Right now, maintaining your family life is extremely important. It's healthy for children to learn that you have goals and that they may have to make compromises some times, but they should never suffer at the expense of those goals. If that means putting off your studies until after they're in bed, then that's what you should do. You might even try studying at the same time they are doing their homework. Make it a family activity. But whatever you do, try and keep your family structure consistent.

My heart goes out to you. You really have a lot on your plate right now. I would also suggest you find some time to care for yourself. I think the greatest stress reducer is exercise. It gets your adrenaline going and keeps your body and mind healthy. You may have to get up a little earlier or work out on your lunch hour like I do, but it's worth the extra effort.

Keep Track of Events

Your date book also enables you to schedule *events*. Rather than thinking of events as separate from goals, tie them to your long-term goals just as you would your other tasks. For example, attending a wedding in a few months contributes to your commitment to spending time with your family. Being aware of quiz dates, due dates for assignments, and meeting dates will aid your goals to achieve in school and become involved.

Note events in your date book so that you can stay aware of them ahead of time. Write them in daily, weekly, monthly, or even yearly sections, where a quick look will remind you that they are approaching. Writing them down will also help you see where they fit in the context of all your other activities. For example, if you have three big tests and a presentation all in one week, you'll want to take time in the weeks before to prepare for them all.

Following are some kinds of events worth noting in your date book:

- Due dates for papers, projects, presentations, and tests
- Important meetings, medical appointments, or due dates for bill payments
- Birthdays, anniversaries, social events, holidays, and other special occasions
- Benchmarks for steps toward a goal, such as due dates for sections of a project or a deadline for losing five pounds on your way to losing twenty

Time Management Strategies

Managing time takes thought and energy. Here are some additional strategies to try.

1. Plan your schedule each week. Before each week starts, note events, goals, and priorities. Look at the map of your week to decide where to fit activities like studying and priority 3 items. For example, if you have a test on Thursday, you can plan study sessions on the days up until then. If you have more free time on Tuesday and Friday than on other days, you can plan workouts or priority 3 activities at those times. Looking at the whole week will help you avoid being surprised by something you had forgotten was coming up.

2. Make and use to-do lists. Use a *to-do list* to record the things you want to accomplish. If you generate a daily or weekly to-do list on a separate piece of paper, you can look at all tasks and goals at once. This will help you consider time frames and priorities. You might want to prioritize your tasks and transfer them to appropriate places in your date book. Some people create daily to-do lists right on their date book pages. You can tailor a to-do list to an important event such as exam week or an especially busy day when you have a family gathering or a presentation to make. This kind of specific to-do list can help you prioritize and accomplish an unusually large task load.

3. Post monthly and yearly calendars at home. Keeping a calendar on the wall will help you stay aware of important events. You can purchase

one or draw it yourself, month by month, on plain paper. Use a yearly or a monthly version (Figure 4-4 shows part of a monthly calendar) and keep it where you can refer to it often. If you live with family or friends, make the calendar a group project so that you stay aware of each other's plans. Knowing each other's schedules can also help you avoid scheduling problems such as two people who need the car at the same time or one partner's scheduling a get-together when the other has to work.

4. Schedule down time. When you're wiped out from too much activity, you don't have the energy to accomplish much with your time. A little down time will refresh you and improve your attitude. Even half an hour a day will help. Fill the time with whatever relaxes you—having a snack, reading, watching TV, playing a game or sport, walking, writing, or just doing nothing. Make down time a priority.

5. Be flexible. Since priorities determine the map of your day, week, month, or year, any priority shift can jumble your schedule. Be ready to reschedule your tasks as your priorities change. On Monday, a homework assignment due in a week might be priority 2. By Saturday, it has become priority 1. On some days a surprise priority such as a medical emergency or a family situation may pop up and force you to cancel everything else on your schedule. Other days a class may be canceled and you will have extra time on your hands. Adjust to whatever each day brings.

> **TERMS**
> **Down time**
> Quiet time set aside for relaxation and low-key activity.

No matter how well you schedule your time, you will have moments when it's hard to stay in control. Knowing how to identify and avoid procrastination and other time traps will help you get back on track.

Figure 4-4 Monthly calendar.

AUGUST 1999

SUNDAY	MONDAY	TUESDAY	WEDNESDAY	THURSDAY	FRIDAY	SATURDAY
1	2 WORK	3 Turn in English paper	4 Dentist 2pm	5 Chem. test	6	7
8 Frank's B-day	9 Psych test WORK	10 6:30 pm Meeting at Student Center	11 Statistics quiz WORK	12 History study group	13 WORK	14 WORK
15	16 WORK	17	18 WORK	19	20	21
22	23	24	25	26	27	28
29	30	31				

Why is Procrastination a Problem?

TERMS

Procrastination The act of putting off something that needs to be done.

Procrastination occurs when you postpone unpleasant or burdensome tasks. People procrastinate for different reasons. Having trouble with goal setting is one reason. People may project goals too far into the future, set unrealistic goals that are too frustrating to reach, or have no goals at all. People also procrastinate because they don't believe in their ability to complete a task or don't believe in themselves in general. As natural as these tendencies are, they can also be extremely harmful. If continued over a period of time, procrastination can develop into a habit that will dominate a person's behavior. Following are some ways to face your tendencies to procrastinate and *just do it!*

Strategies to Fight Procrastination

Weigh the benefits (to you and others) of completing the task versus the effects of procrastinating. What rewards lie ahead if you get it done? What will be the effects if you continue to put it off? Which situation has better effects? Chances are you will benefit more in the long term from facing the task head-on.

Set reasonable goals. Plan your goals carefully, allowing enough time to complete them. Unreasonable goals can be so intimidating that you do nothing at all. "Pay off the credit card bill next month" could throw you. However, "Pay off the credit card bill in six months" might inspire you to take action.

Get started. Going from doing nothing to doing something is often the hardest part of avoiding procrastination. Once you start, you may find it easier to continue.

Break the task into smaller parts. If it seems overwhelming, look at the task in terms of its parts. How can you approach it step-by-step? If you can concentrate on achieving one small goal at a time, the task may become less of a burden. To start, tell yourself, "I only have to read the first two pages, break, and continue."

Ask for help with tasks and projects at school, work, and home. You don't always have to go it alone. For example, if you have put off an intimidating assignment, ask your instructor for guidance. If you need accommodations due to a disability, don't assume that others know about it. Once you identify what's holding you up, see who can help you face the task.

Don't expect perfection. No one is perfect. Most people learn by starting at the beginning and wading through plenty of mistakes and confusion. It's better to try your best than to do nothing at all.

Procrastination is natural, but it can cause you problems if you let it get the best of you. When it does happen, take some time to think about the causes. What is it about this situation that frightens you or puts you off? Answering that question can help you address what causes lie underneath the procrastination. These causes might indicate a deeper problem that needs to be solved.

In Hebrew, the word *chai* means "life," representing all aspects of life—spiritual, emotional, family, educational, and career. Individual Hebrew characters have number values. Because the characters in the word *chai* add up to 18, the number 18 has come to be associated with good luck. The word *chai* is often worn as a good luck charm. As you plan your goals, think about your view of luck. Many people feel that a person can create his or her own luck by pursuing goals persistently and staying open to possibilities and opportunities. Canadian novelist Robertson Davies once said, "What we call luck is the inner man externalized. We make things happen to us."

Consider that your vision of life may largely determine how you live. You can prepare the way for luck by establishing a personal mission and forging ahead toward your goals. If you believe that the life you want awaits you, you will be able to recognize and make the most of luck when it comes around. *L'chaim*—to life, and good luck.

Chapter 4   Applications

Name _____   Date _____

KEY INTO YOUR LIFE
Opportunities to Apply What You Learn

 ### 4.1 *Your Values*

Begin to explore your values by rating the following values on a scale from 1 to 4, 1 being least important to you and 4 being most important. If you have values that you don't see in the chart, list them in the blank spaces and rate them.

VALUE	RATING	VALUE	RATING
Knowing yourself		Mental health	
Physical health		Fitness and exercise	
Spending time with your family		Having an intimate relationship	
Helping others		Education	
Being well paid		Being employed	
Being liked by others		Free time/vacations	
Enjoying entertainment		Time to yourself	
Spiritual/religious life		Reading	
Keeping up with the news		Staying organized	
Being financially stable		Close friendships	
Creative/artistic pursuits		Self-improvement	
Lifelong learning		Facing your fears	

Considering your priorities, write your top five values here:

1. _____
2. _____
3. _____
4. _____
5. _____

4.2 Why Are You Here?

Why did you decide to enroll in school? Do any of the reasons listed in the chapter fit you? Do you have other reasons all your own? Many people have more than one answer. Write up to five here.

_____
_____
_____
_____
_____

Take a moment to think about your reasons. Which reasons are most important to you? Why? Prioritize your reasons above by writing 1 next to the most important, 2 next to the second most important, etc.

How do you feel about your reasons? You may be proud of some. On the other hand, you may not feel comfortable with others. Which do you like or dislike and why?

_____
_____
_____
_____
_____

4.3 Short-Term Scheduling

Take a close look at your schedule for the coming month, including events, important dates, and steps toward goals. On the calendar layout on p. 103, fill in the name of the month and appropriate numbers for the days. Then record what you hope to accomplish, including the following:

- Due dates for papers, projects, and presentations
- Test dates
- Important meetings, medical appointments, and due dates for bill payments
- Birthdays, anniversaries, and other special occasions
- Steps toward long-term goals

This kind of chart will help you see the big picture of your month. To stay on target from day to day, check these dates against the entries in your date book and make sure that they are indicated there as well.

 ## To-Do Lists

Make a to-do list for what you have to do tomorrow. Include all tasks—priority 1, 2, and 3—and events.

TOMORROW'S DATE: _____

1. _____
2. _____
3. _____
4. _____
5. _____
6. _____
7. _____
8. _____
9. _____
10. _____

Use a coding system of your choice to indicate priority level of both tasks and events. Place a check mark by the items that are important enough to note in your date book. Use this list to make your schedule for tomorrow in the date book, making a separate list for priority 3 items. At the end of the day, evaluate this system. Did the to-do list help you? How did it make a difference? If you liked it, use this exercise as a guide for using to-do lists regularly.

_____
_____
_____
_____

September

MONTH CHART

	1	2	3	4	5
6 study day)	7	8 car payment	9 license expires. Go to DMV.	10	11 payday	12 Brothers soccer Game 10:30am.
13	14	15	16 HLT 141 Test #1	17	18	19 power-point presentation.
20	21	22	23 Dentist appt	24 Sose's B-day	25	26 career presentation
27	28	29 Angel's Birthday	30			

KEY TO SELF-EXPRESSION
Discovery Through Journal Writing

To record your thoughts, use a separate journal or the lined pages at the end of the chapter.

Using the personal mission statement examples in the chapter as a guide, consider what you want out of your life and create your own personal mission statement. You can write it in paragraph form, in a list of long-term goals, or in the form of a think link. Take as much time as you need in order to be as complete as possible.

Name _____ Date _____ **Journal**

Journal

Name _____ Date _____

Scientific Inquiry

Critical and Creative Thinking

Your mind's powers show in everything you do, from the smallest chores (comparing prices on cereals at the grocery store) to the most complex situations (figuring out how to earn money after being laid off). Your mind is able to process, store, and create with the facts and ideas it encounters. Critical and creative thinking are what enable those skills to come alive.

Understanding how your mind works is the first step toward critical thinking. When you have that understanding, you can perform the essential critical-thinking task: asking important questions about ideas and information. This chapter will show you both the mind's basic actions and the thinking processes that use those actions. You will explore what it means to be an open-minded critical and creative thinker able to ask and understand questions that promote your success as a science major, in your career, and in life.

In this chapter, you will explore answers to the following questions:

- What is critical thinking?
- How is critical thinking critical in science?
- How does your mind work?
- How does critical thinking help you solve problems and make decisions?
- Why shift your perspective?
- Why plan strategically?
- How can you develop creativity in science?

WHAT IS CRITICAL THINKING?

Critical thinking is thinking that goes beyond the basic recall of information. If the word *critical* sounds negative to you, consider that the dictionary defines its meaning as "indispensable" and "important." Critical thinking is important thinking that involves asking questions. This is called essential questioning. Using critical thinking, you question established ideas, create new ideas, turn information into tools to solve problems and make decisions, and take the long-term view as well as the day-to-day view.

A critical thinker asks as many kinds of questions as possible. The following are examples of possible questions about a given piece of information: *Where did it come from? What could explain it? In what ways is it true or false, and what examples could prove or disprove it? How do I feel about it, and why? How is this information similar to or different from what I already know? Is it good or bad? What causes led to it, and what effects does it have?* Critical thinkers also try to transform information into something they can use. They ask themselves whether the information can help them solve a problem, make a decision, create something new, or anticipate the future. Such questions help the critical thinker learn, grow, and create.

Not thinking critically means not asking questions about information or ideas. A person who does not think critically tends to accept or reject information or ideas without examining them. Table 5-1 compares how a critical thinker and a non-critical thinker might respond to particular situations.

Asking questions (the focus of the table), considering without judgment as many responses as you can, choosing responses that are as complete and accurate as possible, and having insight into your own biases are some primary ingredients that make up the skill of critical thinking. You must be willing to ask questions.

Critical Thinking Is a Skill

Anyone can develop the ability to think critically. Critical thinking is a skill that can be taught to students at all different levels of ability. One of the most crucial components of this skill is learning information. For instance, part of

Table 5-1 Not thinking critically vs. thinking critically.

YOUR ROLE	SITUATION	NON-QUESTIONING RESPONSE	QUESTIONING RESPONSE
STUDENT	Instructor is lecturing on the causes of congestive heart failure.	You assume that everything your instructor tells you is true.	You consider what the instructor says; you write down questions about issues you want to clarify; you initiate discussion with the professor or other classmates.
PARENT	Teacher discovers your child lying about something at school.	You're mad at your child and believe the teacher, or you think the teacher is lying.	You ask both teacher and child about what happened, and you compare their answers, evaluating who you think is telling the truth; you discuss the concepts of lying/honesty with your child.
SPOUSE/ PARTNER	Your partner feels that he or she no longer has quality time with you.	You think he or she is wrong and defend yourself.	You ask how long he or she has felt this way; you ask your partner and yourself why this is happening; you explore how you can improve the situation.
EMPLOYEE	You are angry at your supervisor.	You ignore or avoid your supervisor.	You are willing to discuss the situation.
NEIGHBOR	People different from you move in next door.	You ignore or avoid them; you think their way of living is weird.	You introduce yourself; you offer to help if they need it; you respectfully explore what's different about them.
CITIZEN	You encounter a homeless person.	You avoid the person and the issue.	You examine whether the community has a responsibility to the homeless, and if you find that it does, you explore how to fulfill that responsibility.
CONSUMER	You want to buy a car.	You decide on a brand-new car and don't think through how you will handle the payments.	You consider the different effects of buying a new car vs. buying a used car; you examine your money situation to see what kind of payment you can handle each month.

the skill of critical thinking is comparing new information with what you already know. Your prior knowledge provides a framework within which to ask questions about and evaluate a new piece of information. Without a solid base of knowledge, critical thinking is harder to achieve. For example, thinking critically about the statement "Shakespeare's character King Richard III is like an early version of Adolf Hitler" is impossible without basic knowledge of World War II and Shakespeare's play *Richard III*.

The skill of critical thinking focuses on generating questions about statements and information. To examine potential critical-thinking responses in more depth, explore the different questions that a critical thinker may have about one particular statement.

A Critical-Thinking Response to a Statement

Consider the following statement of opinion: *"My obstacles are keeping me from succeeding in school. Other people make it through school because they don't have to deal with the obstacles that I have."*

Non-questioning thinkers may accept an opinion such as this as an absolute truth, believing that their obstacles will hinder their success. As a result, on the road to achieving their goals, they may lose motivation to overcome those obstacles. In contrast, critical thinkers would take the opportunity to examine the opinion through a series of questions. Here are some examples of questions one student might ask.

> *"What exactly are my obstacles?* Examples of my obstacles are a heavy work schedule, single parenting, being in debt, and returning to school after 10 years out."
>
> *"Who has problems that are different from mine?* I do have one friend who is going through problems worse than mine, and she's getting by. I also know another guy who doesn't have too much to deal with that I can tell, and he's struggling just like I am."
>
> *"Who has problems that are similar to mine?* Well, if I consider my obstacles specifically, I might be saying that single parents and returning adult students will all have trouble in school. That is not necessarily true. People in all kinds of situations may still become successful."
>
> *"Why do I think this?* Maybe I am scared of returning to school and adjusting to a new environment. Maybe I am afraid to challenge myself, which I haven't done in a long time. Whatever the cause, the effect is that I feel bad about myself and don't work to the best of my abilities, and that can hurt both me and my family who depends on me."
>
> *"What is an example of someone who has had success despite having to overcome obstacles?* What about Oseola McCarty, the cleaning woman who saved money all her life and raised $150,000 to create a scholarship at the University of Southern Mississippi? She didn't have what anyone would call advantages, such as a high-paying job or a college education."
>
> *"What conclusion can I draw from my questions?* From thinking about my friend and about Oseola McCarty, I would say that people can successfully overcome their obstacles by working hard, focusing on their abilities, and concentrating on their goals."

"How do I evaluate the effects of my statement? I think it's harmful. When we say that obstacles equal difficulty, we can damage our desire to try to overcome those obstacles. When we say that successful people don't have obstacles, we might overlook that some very successful people have to deal with hidden disadvantages such as learning disabilities or abusive families."

The Value of Critical Thinking

Critical thinking has many important advantages. Following are some of the positive effects, or benefits, of putting energy into critical thinking.

You will increase your ability to perform thinking processes that help you reach your goals in science. Critical thinking is a learned skill, just like shooting a basketball or using a new software program on the computer. As with any other skill, the more you use it, the better you become. The more you ask questions, the better thinker you become. The better you think, the more effective you will be when learning in school, managing your personal life, and performing and being valued on the job.

You can produce knowledge, rather than just reproduce it. The interaction of new information with what you already know creates new knowledge. The usefulness of such knowledge can be judged by how you apply it to new situations. For instance, it won't mean much for a pharmacy major to list the major drug categories on an exam unless he or she can make judgments about drug interactions when on the job.

You can be a valuable employee. You certainly won't be a failure in the workplace if you follow directions. However, you will be even more valuable if you can think critically and ask strategic questions about how to make improvements, large or small. Questions could range from "Is there a better way to deliver phone messages?" to "How can we find new grant money to keep our research project going?" An employee who shows the initiative to think critically will be more likely to earn responsibility and promotions.

You can increase your creativity. You cannot be a successful critical thinker without being able to come up with new and different questions to ask, possibilities to explore, and ideas to try. Creativity is essential in producing what is new. Being creative generally improves your outlook, your sense of humor, and your perspective as you cope with problems. Later in this chapter, you will look at ways to awaken and increase your natural creativity.

"We do not live to think but, on the contrary, we think in order that we may succeed in surviving."

JOSÉ ORTEGA Y GASSETT

HOW IS CRITICAL THINKING CRITICAL IN SCIENCE?

Critical thinking in science is a process of inquiry that tries to gain a better understanding of the world—from stars and meteors to computers and suspension bridges to entire ecosystems. Inquiry is based on a standard set of rules known as the scientific method. The scientific method is important because it provides a regulated process for conducting research:

- Essential questioning: asking questions
- Possible answers: forming hypotheses
- Testing hypotheses: looking for answers

The scientific method is a process that other scientists can then follow and repeat to reproduce and validate your results. Repetition of research studies gives the results more strength by increasing the amount of supporting evidence. For instance, you'd like to know that a medication you took to fight a bacterial infection had been researched using a standard method, tested repeatedly, and had strong evidence supporting its effectiveness. Furthermore, you would want to be confident that it worked on the specific bacteria you had and that it didn't have any dangerous side effects.

The main ingredient of the scientific method is the ability to think, which sounds pretty easy, perhaps like breathing. But, you can learn to improve your thinking as you progress through college course work. Even thinking about your own thinking, called reflection, can help you. Reflection helps you understand your own biases, or your particular way of looking at phenomena, so that you can find out how your previous views might be getting in the way when what you need is a fresh perspective.

Observation is a critical skill in inquiry, and you can learn to become an astute observer through practicing the journal exercises in Chapters 1 and 2. Another thinking skill you can learn is making connections between what you already know and what you are learning. This skill will help you put information together to make new discoveries or to come up with new solutions to old problems.

Inquiry in science relies on asking critical questions. Questions help direct your inquiry; they help you decide where to go for information, what tests to perform, or what experiments to design. The more you improve your thinking through practice and experience, the better you will be at coming up with questions about the world, or your area of study, and finding methods for answering those questions.

In the next section, you will read about the seven basic actions your mind performs when asking important questions. These actions are the basic blocks you will use to build the critical-thinking processes you will explore later in the chapter.

How Does Your Mind Work?

Critical thinking depends on a thorough understanding of the workings of the mind. Your mind has some basic moves, or actions, that it uses to understand relationships among ideas and concepts. Sometimes it uses one action by itself, but most often it uses two or more in combination.

Brain research has advanced rapidly in the past decade. The discovery of new technologies, or new uses for old technologies, has led to discoveries about memory, moods, and the learning process. For instance, researchers have used brain-scanning technology to show how parts of the brain respond to depression. This in turn has helped other researchers develop medications that work on a cellular level in brain tissue to effectively treat depression.

Brain-scanning techniques are also used to study a person's brain while they are learning. For instance, while learning to play the piano, brain scans found that the brain recruited other areas of the cerebral cortex to help in the learning process. Once the learning was complete and the function became automatic, like learning to ride a bicycle, those borrowed areas went back to their normal functions. This helps to explain how the mind works to learn new information.

Mind Actions: The Thinktrix

You can identify your mind's actions using a system called the Thinktrix, originally conceived by educators Frank Lyman, Arlene Mindus, and Charlene Lopez[1] and developed by numerous other instructors. They studied how students think and named seven mind actions that are the basic building blocks of thought. These actions are not new to you, although some of their names may be. They represent the ways in which you think all the time.

Through exploring these actions, you can go beyond just thinking and learn *how* you think. This will help you take charge of your own thinking. The more you know about how your mind works, the more control you will have over thinking processes such as problem solving, decision making, creating, and strategic planning.

Following are explanations of each of the mind actions. Each explanation names the action, defines it, and explains it with examples. As you read, follow the instructions and write your own examples in the blank spaces provided. Each action is also represented by a picture, or *icon*, that helps you visualize and remember it.

Recall: *Facts, sequence, and description.* This is the simplest action. When you **recall** you describe facts, objects, or events, or put them into sequence. *Examples:*

- Naming the steps of a geometry proof, in order
- Remembering the valence electron arrangement for calcium, chloride, and nitrogen

Your example: Recall some important events this month. _____
_____
_____
_____

The icon: A string tied around a finger is a familiar image of recall or remembering.

Similarity: *Analogy, likeness, comparison.* This action examines what is **similar** about one or more things. You might compare situations, ideas, people, stories, events, or objects. *Examples:*

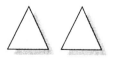

- Comparing notes with another student to see what facts and ideas you both have considered important.
- Analyzing the arguments you've had with your partner this month and seeing how they all seem to be about the same problem.

Your example: Tell what is similar about the numbers 2, 4, 8, 12, and 16. _____

_____
_____
_____

The icon: Two alike objects, in this case triangles, indicate similarity.

Difference: *Distinction, contrast, comparison.* This action examines what is **different** about one or more situations, ideas, people, stories, events, or objects, contrasting them with one another. *Examples:*

- Seeing how three mammals differ in anatomy (horse, human, leopard)
- Contrasting a weekday where you work half the day and go to school half the day with a weekday when you attend class and then have the rest of the day to study

Your example: Explain what about one course makes it easier to learn the material than in another course. _____

_____
_____
_____

The icon: Two differing objects, in this case a triangle and a square, indicate difference.

Cause and effect: *Reasons, consequences, prediction.* Using this action, you look at what has **caused** a situation, or event, and what **effects,** or consequences, come from it. In other words, you examine both what led up to something and what will follow because of it. *Examples:*

- You see how staying up late at night causes you to oversleep, which has the effect of your being late to class. This causes you to miss some of the material, which has the further effect of your having problems on the test.
- When you pay your phone and utility bills on time you create effects such as a better credit rating, uninterrupted service, and a better relationship with your service providers.

Your example: Name probable causes for the days' becoming shorter in winter.

_____
_____

The icon: The water droplets making ripples indicate causes and their resulting effects.

Example to principle: *Inductive reasoning, generalization, classification, conceptualization.* From one or more **examples** (facts or events), you develop a general **principle** or principles. Grouping facts or events into patterns may allow you to make a general statement about several of them at once. Classifying a fact or event helps you build knowledge. This mind action moves from the specific to the general. *Examples:*

- Molecules such as hydrochloric acid, sulfuric acid, and acetic acid all easily lose a proton (examples). Therefore, all molecules with a certain structure are an acid (principle).
- You see a movie (example), and you decide it is mostly about pride (principle).

Your example: You drop a pencil and it falls down; you see an apple fall from a tree.
Name the principle: _____

_____

_____

_____

The icon: The arrow and "Ex" pointing to a light bulb on their right indicate how an example or examples lead to the principle, or idea (the light bulb, lit up).

Principle to Example: *Deductive reasoning, categorization, substantiation, proof.* In a reverse of the previous action, you take a **principle** or principles and think of **examples** (events or facts) that support or prove that idea. This mind action moves from the general to the specific. *Examples:*

- When you write a paper, you start with a thesis statement, which communicates the central idea: "Men are favored over women in the modern workplace." Then you gather examples to back up that idea: Men make more money on average than women in the same jobs; there are more men in upper management positions than there are women; women can be denied advancement when they make their families a priority.
- You talk to your instructor about changing your major, giving examples that support your idea: You have worked in the field you want to change to, you have fulfilled some of the requirements for that major already, and you are unhappy with your current course of study.

Your example: Air pressure changes with altitude changes (principle). Name an example you might experience that supports the principle. _____

_____

_____

The icon: In a reverse of the previous icon, this one starts with the light bulb and has an arrow pointing to "Ex." This indicates that you start with the principle, or idea, the light bulb, and then branch into the supporting example or examples.

Evaluation: *Analysis, value, judgment, rating.* Here you **judge** whether something is useful or not useful, important or unimportant, good or bad, or right or wrong by identifying and weighing its positive and negative effects (pros and cons). Be sure to consider the specific situation at hand (a cold drink might be good on the beach in August, but not so good in the snowdrifts in January). With the facts you have gathered, you determine the value of something in terms of both predicted effects and your own needs. Cause and effect analysis always accompanies evaluation. *Examples:*

- You decide to try taking classes later in the day for a semester. You schedule classes in the afternoons and spend your nights on the job. You find that instead of getting up early to use the morning time, you tend to sleep in and then get up not too long before you have to be at school. From those harmful effects, you evaluate that it doesn't work for you. You decide to schedule earlier classes next time.

- Someone offers you a chance to cheat on a test. You evaluate the potential effects if you are caught. You also evaluate the long-term effects on you of not actually learning the material. You decide that it isn't worth your while to participate in the plan to cheat.

Your example: Evaluate your mode of transportation to school. _____

_____

_____

The icon: A set of scales out of balance indicates how you weigh positive and negative effects to arrive at an evaluation.

You may want to use a *mnemonic device*—a memory tool, explained in more detail in Chapter 8—to remember the seven mind actions. You can make a sentence of words that each start with a mind action's first letter. Here's an example: "Really Smart Dogs Cook Enchiladas Producing Energy" (the first letter of each word stands for one of the mind actions).

How Mind Actions Build Thinking Processes

The seven mind actions are the fundamental building blocks that your mind uses every day. Note that you will rarely use them one at a time in a step-by-step process, as they are presented here. You will usually combine them, overlap them, and repeat them more than once, using different actions for different situations. For example, when you want to say something nice at the end of a date, you might consider past comments that had an effect *similar* to what you want now. When a test question asks you to explain what prejudice is, you might name similar *examples* that confirm a principle, or theory, on prejudice.

When you combine mind actions in working toward a specific goal, you are performing a thinking process. The next few sections will explore some of the most important critical-thinking processes: solving problems, making decisions, shifting your perspective, and planning strategically. Each thinking process helps you succeed by directing your critical thinking toward the achievement of your goals. Figure 5-4, appearing later in the chapter, shows all of the mind actions and thinking processes together and reminds you that the mind actions form the core of the thinking processes.

HOW DOES CRITICAL THINKING HELP YOU SOLVE PROBLEMS AND MAKE DECISIONS?

Problem solving and decision making are probably the two most crucial and common thinking processes used in science. Each one requires various mind actions. They overlap somewhat, because every problem that needs solving requires you to make a decision. Each process will be considered separately here. You will notice similarities in the steps involved in each.

Although both of these processes have multiple steps, you will not always have to work your way through each step. As you become more comfortable with solving problems and making decisions, your mind will automatically click through the steps you need whenever you encounter a problem or decision. Also, you will become more adept at evaluating which problems and decisions need serious consideration and which can be taken care of more quickly and simply.

Problem Solving and Inquiry

Problem solving starts with asking questions. Asking questions is the fundamental step in inquiry. Problems to solve can range from very small, such as how to learn not to lose your car keys, to very large, such as how to care for a sick relative, manage a custody plan after a divorce, or protect the ozone layer. A problem, however, is not always necessarily a problem, that is, something that has to be fixed. In science, a problem is something to investigate and gain understanding about. For instance, the problem I want to learn more about is how people respond to aerobic exercise that is alternated with short bursts of anaerobic exercise.

Solving a problem can occur quickly or over many years of investigation. How quickly a solution is reached may depend on two things: (1) The urgency of the problem; and (2) the complexity of the problem.

Solving the problem of the AIDS virus is extremely urgent; it is a problem that affects people of all ages and all over the world. It is also a complex problem. Another complex problem, but one that may be judged as less urgent, is finding out how the universe began. An urgent problem requiring a quick solution is an emergency. For instance, if you have the signs and symptoms of a heart attack while playing soccer, you have an urgent, but fairly simple problem (although it may be complex in the long run). The solution doesn't

"Most of our knowledge, and animal knowledge, and even vegetable knowledge, is rather the result of sheer invention. All organisms are professional problem solvers: before life, problems did not exist. Problems and life entered the world together and with them problem solving."

KARL POPPER

require much critical thinking or take much time to come up with. Someone calls 911, and you go to the emergency department of the nearest hospital. More complex problems require more complex thinking, or critical thinking.

Use the following steps to generate ideas, questions, and testable hypotheses:

1. Observe

State the problem clearly. To do this, figure out what you already know about the problem by noting what objective, or observable, evidence you have. If you are failing physics quizzes, you may think the problem is that you don't understand the material. Not understanding the material is not the problem; it is a possible *cause* of the observable problem, which is your failing grades.

2. Assess and Analyze

After the problem is defined, "I am failing my physics quizzes," it is time to assess what else you need to know in order to develop a solution. This is when you formulate questions. In this case you might ask yourself: "What are possible causes of my poor grades—lack of sleep or too much TV watching?" And to determine the urgency of the problem: "What is the consequence of failing these quizzes?" "Will I fail the class?" "Will my plan to go into biomedical engineering be ruined?" "Will I ever be happy again?"

3. Brainstorm

Brainstorming helps loosen up your thinking and creativity. Brainstorming is done without judging your answers, or ideas; thus, you remove a major barrier to creativity: fear that your answer will be wrong. That is the beauty of brainstorming—there are no right or wrong answers. This gives you the freedom to let your mind come up with all kinds of possibilities, some useful, some useless. Again, ask yourself: "What are possible causes of my failing grades?" Now, think up every cause you can, focusing just on causes, not effects or solutions.

4. Plan

Explore possible solutions. This stage is like the scientific process stage of coming up with testable hypotheses. For instance, you think that lack of sleep may be the biggest cause of your misunderstanding the physics material and hence, the problem—failing the quizzes. One hypothesis: "Increasing sleep at night to at least eight hours will increase alertness and receptiveness to understanding the physics material more than four to five hours of sleep per night did. Now you can test the hypothetical solution: getting more sleep. If this doesn't work (except to make you feel better, which counts, too), you may test a second hypothesis: "A tutor will improve grades."

5. Implement

Choose and implement the solution you think is most probable in improving your grades.

TERMS

Brainstorming
The spontaneous, rapid generation of ideas or solutions, undertaken by a group or an individual, often as part of a problem-solving process.

6. Evaluate

Evaluate the results you get from testing your hypothesis. In the physics student's case, did getting more sleep or working with a tutor help her grades? Look at the pros and cons of the solution. Getting eight hours of sleep may make you feel better—a pro, but it didn't help your grades—a con. Using a tutor helped your grades—a pro, but it took up time you wanted to use for skiing or other activities—a con. You must weigh the pros against the cons. Less skiing and good grades today will help you find a job later that you enjoy and that pays well enough for you to enjoy plenty of skiing later.

7. Refine

Problem solving is not linear; it is circular. There is no definite beginning or end. As you work on a problem, new information will arise. Perhaps the method or instruments you are using do not measure what you want to measure. Then you will rethink your problem and method and possibly change your instrument. Refining means rethinking. It means weighing the pros and cons of your method, or of your solutions. The physics student may decide to continue to study regularly, but after several weeks, to drop the tutoring. Although the original plan was to use the tutor for longer, the student found it possible to apply, on her own, what was learned from the tutor. Thus, more time for other activities was now possible.

Using this process will enable you to solve personal, educational, and workplace problems in a thoughtful, comprehensive way. Figure 5-1 is a think link that demonstrates a way to visualize the flow of problem solving. Figure 5-2 contains a sample of how one person used this plan to solve a problem. Figure 5-3 leaves space for writing so that it can be used in your problem-solving process.

Decision Making

Decisions are choices. Although every problem-solving process involves making a decision (when you decide which solution to try), not all decisions involve solving problems. Making a choice, or decision, requires thinking critically through all of the possible choices and evaluating which will work best for you and for the situation. Decisions large and small come up daily, hourly, even every few minutes. Do you call your landlord when the heat isn't coming on? Do you drop a course? Should you stay in a relationship? Can you work part-time without interfering with school?

Before you begin the decision-making process, evaluate the level of the decision you are making. Do you have to decide what to have for lunch (usually a minor issue), or whether to quit a good job (often a major life change)? Some decisions are little day-to-day considerations that you can take care of quickly on your own. Others require thoughtful evaluation, time, and perhaps the input of others you trust. The following is a list of steps to take in order to think critically through a decision:

1. Decide on a goal. Why is this decision necessary? In other words, what result do you want from this decision, and what is its value?

Figure 5-1

Problem-solving plan.

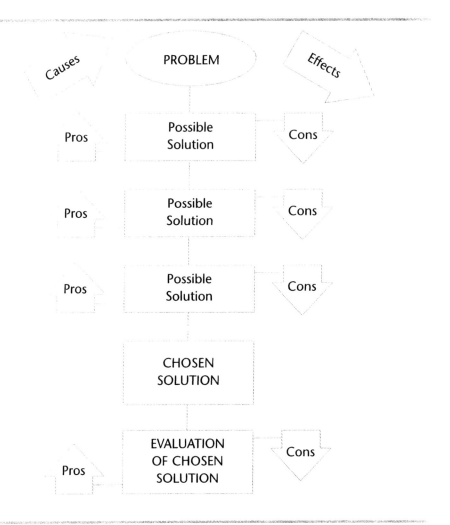

Considering the *effects* you want can help you formulate your goal. For example, say a student currently attends a small private college. Her goal is to become a physical therapist. The school has a good program, but her financial situation has changed and has made this school too expensive for her.

2. Establish needs. *Recall* the needs of everyone (or everything) involved in the decision. The student needs a school with a full physical therapy program; she and her parents need to cut costs (her father changed jobs and her family cannot continue to afford the current school); she needs to be able to transfer credits.

3. Name, investigate, and evaluate available options. Brainstorm possible choices, and then look at the facts surrounding each. *Evaluate* the good and bad effects of each possibility. Weigh these effects, and judge which is the best course of action. Here are some possibilities that the student in the college example might consider:

   ◆ *Continue at the current college.* **Positive effects:** I wouldn't have to adjust to a new place or to new people. I could continue my course work as planned. **Negative effects:** I would have to find a way to finance most of my tuition and costs on my own, whether through

Figure 5-2 How one student worked through a problem.

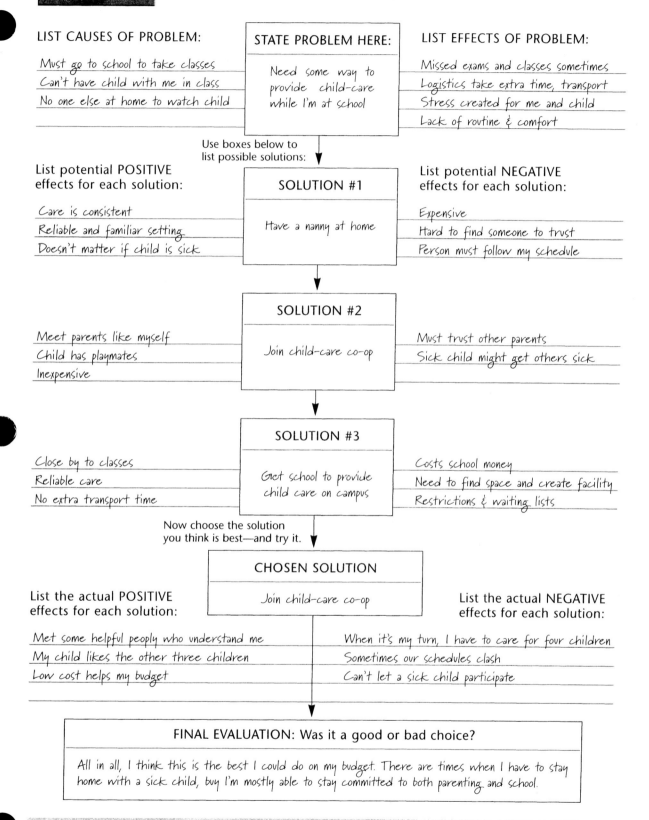

Source: Adapted from a heuristic developed by Frank T. Lyman Jr., Ph.D., University of Maryland, 1983.

122 CHAPTER 5 Scientific Inquiry

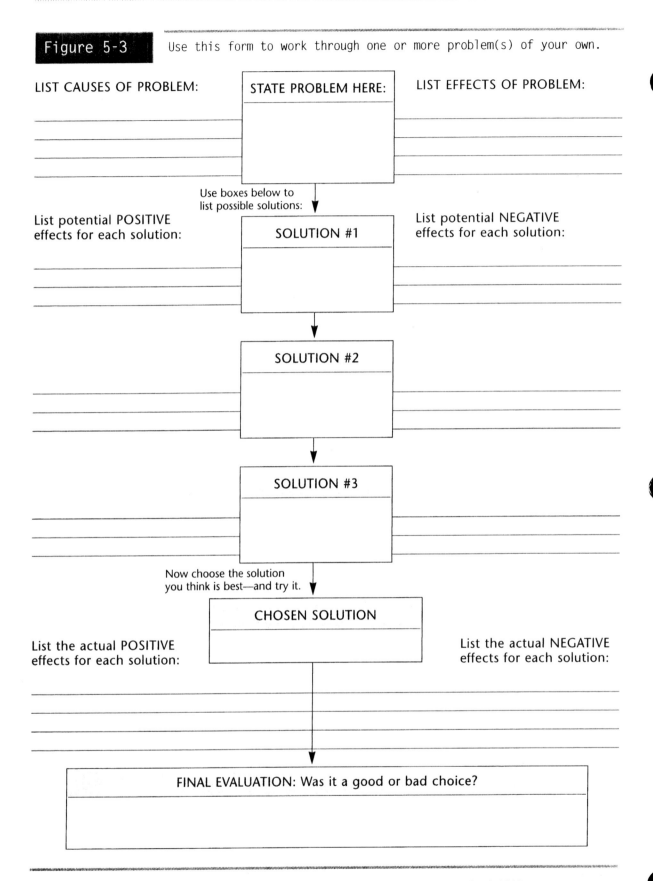

Figure 5-3 Use this form to work through one or more problem(s) of your own.

Source: Adapted from a heuristic developed by Frank T. Lyman Jr., Ph.D., University of Maryland, 1983.

loans, grants, or work. I'm not sure I could find time to work as much as I would need to, and I don't think I would qualify for as much aid as I now need.

- *Transfer to the state college.* **Positive effects:** I could reconnect with people there that I know from high school. Tuition and room costs would be cheaper than at my current school. I could transfer credits. **Negative effects:** I would still have to work some or find minimal financial aid. The physical therapy program is small and not very strong.

- *Transfer to the community college.* **Positive effects:** They have many of the courses I need to continue with the physical therapy curriculum. The school is 20 minutes from my parents' house, so I could live at home and avoid paying housing costs. Credits will transfer. The tuition is extremely reasonable. **Negative effects:** I don't know anyone there. I would be less independent. The school doesn't offer a bachelor's degree.

4. Decide on a plan of action and pursue it. Make a choice based on your evaluation, and act on your choice. In this case the student might decide to go to the community college for two years and then transfer back to a four-year school to earn a bachelor's degree in physical therapy. Although she might lose some independence and contact with friends, the positive effects are money saved, opportunity to spend time on studies rather than working to earn tuition money, and the availability of classes that match the physical therapy program requirements.

5. Evaluate the result. Was it useful? Not useful? Some of both? Weigh the positive and negative effects. The student may find with her transfer decision that it can be hard living at home, although her parents are adjusting to her independence and she is trying to respect their concerns as parents. Fewer social distractions result in her getting more work done. The financial situation is much more favorable. All things considered, she evaluates that this decision was a good one.

Making important decisions can take time. Think through your decision thoroughly, considering your own ideas as well as those of others you trust, but don't hesitate to act once you have your plan. You cannot benefit from your decision until you act upon it and follow through.

WHY SHIFT YOUR PERSPECTIVE?

Seeing the world from only your perspective, or point of view, is inflexible, limiting, and frustrating to both you and others. You probably know how hard it can be to relate to someone who cannot understand where you are coming from—a co-worker who's annoyed that you leave early on Thursdays for physical therapy, a parent who doesn't see why you can't take a study break to visit, a friend who can't understand why you would date someone of a different race. Seeing beyond one's own perspective can be difficult, especially when life problems and fatigue take their toll.

Perspective
A mental point of view or outlook, based on a cluster of related assumptions, incorporating values, interests, and knowledge.

REAL WORLD PERSPECTIVE

How can I find a satisfactory solution to my problem?

Chelsea Phillips, Hampshire College, Massachusetts, Environmental Science Major

I attend Hampshire College in Massachusetts. This year I'm involved in a field study program called Earth Lands. The college gives me credits, but the program is not affiliated with Hampshire. I live and work in a sustainable community and study ecological issues. There are nine of us that live together. All of us are environmental activists, and we agree to live by certain principles. The lodge we live in is run by solar power. We use kerosene and flashlights, too. Our food is entirely vegan, which means we not only don't eat meat, we also don't eat other foods that come from animals, like milk and butter.

Five of the participants in the program, including myself, are here as paying students. The other members are brought in to live with us and support us as we learn about the environment and community living. When we got involved, we believed the program was an entirely collaborative effort—at least that's what the brochure said. We're coming to find out there is a subtle power structure that exists between the five of us and the group called the "centering Team." We don't have as much input as we'd like into the schedule or decisions that need to be made. Because we're learning how to build a community and resolve problems, I'd like to find a way to resolve this feeling of separation between the two groups. I'd like to see much more dialogue and collaboration so that we're all equal participants. What process could I initiate that would address this problem and allow for more equality within our community?

Raymond Reyes, Community and Organizational Consultant

There seems to be a "tale of two cities" where there are two distinct groups of people. I would recommend that you revisit and "reclaim" the core principles that you have said were agreed upon by everyone in the community. There is an obvious gap between what has been said and reality. As a community, you need to journey into the gap, or what Plato called, "the fertile void." You may want to give serious consideration to identifying and inviting an individual who can guide you through a process to establish a greater level of trust and authenticity and to do some team building.

Communities and other "learning organizations" need to address what I often refer to as the "other three R's" of education: relationship, relevance, and respect. First, address the need for honest and healthy relationships by specifically identifying and working through the trust and power issues. Second, make the core principles upon which your community is based more relevant so that the members truly "own" them whether they are paying students or part of the "centering Team." Last, your community needs to establish a social culture that has "wake-up" calls that remind everyone to practice respect. Just as you are practicing respect for our Earth Mother, your community needs to have the daily fellowship behaviors that are likewise respectful.

On the other hand, when you shift your own perspective to consider someone else's, you open the lines of communication. Trying to understand what other people feel, need, and want makes you more responsive to them. They then may feel respected by you and respond to you in turn. For example, if you want to add or drop a course and your advisor says it's impossible, not waiting to hear you out, the last thing you may feel like doing is pouring your heart out. On the other hand, if your advisor asks to hear your point of view, you may sense that your needs are respected. Because the advisor wants to hear from you, you feel valued; that may encourage you to respond, or even to change your mind.

Every time you shift your perspective, you can also learn something new. There are worlds of knowledge and possibilities outside your individual existence. You may learn that what you eat daily may be against someone else's religious beliefs. You may discover people who don't fit a stereotype.

Shifting your perspective is invaluable in helping you find fresh solutions. In science, a fresh view is extremely useful in enhancing creativity.

Asking questions like these will help you maintain flexibility and openness in your perspective.

- What is similar and different about this person/belief/method and me/my beliefs/my methods?

- What positive and negative effects come from this different way of being/acting/believing? Even if this perspective seems to have negative effects for me, how might it have positive effects for others and therefore have value?

- What can I learn from this different perspective? Is there anything I could adopt for my own life—something that would help me improve who I am or what I do? Is there anything I wouldn't do myself but that I can still respect and learn from?

Shifting your perspective is at the heart of all successful communication. Each person is unique. Even within a group of people similar to yourself, there will be a great variety of perspectives. Whether you see that each world community has different customs or you understand that a friend can't go out on weekends because he spends that time with his mother, you have increased your wealth of knowledge and shown respect to others. Being able to shift perspective and communicate more effectively may mean the difference between success and failure in today's diverse working world.

WHY PLAN STRATEGICALLY?

If you've ever played a game of chess or checkers, participated in a wrestling or martial arts match, or had a drawn-out argument, you have had experience with strategy. In those situations and many others, you continually have to think through and anticipate the moves the other person is about to make. Often you have to think about several possible options that person could put into play, and you consider what you would counter with should any of those options occur. In competitive situations, you try to outguess the other person with your choices. The extent of your strategic skills can determine whether you will win or lose.

Strategy is the plan of action, the method, the "how" behind any goal you want to achieve. Specifically, strategic planning means having a plan for the future, whether you are looking at the next week, month, year, 10 years, or 50

TERMS

Strategy
A plan of action designed to accomplish a specific goal.

years. It means exploring the future positive and negative effects of the choices you make and actions you take today. You are planning strategically right now just by being in school. You made a decision that the requirements of attending college are a legitimate price to pay for the skills, contacts, and opportunities that will help you in the future. As a student, you are challenging yourself to achieve. You are learning to set goals for the future, analyze what you want in the long term, and prepare for the job market to increase your career options. Being strategic with yourself means challenging yourself as you would challenge a competitor, urging yourself to work toward your goals with conviction and determination.

What are the benefits, or positive effects, of strategic planning?

Strategy is an essential skill in the workplace. A food company that wants to develop a successful health-food product needs to examine the anticipated trends in health consciousness. A nurse practitioner needs to think through every aspect of the client's case, anticipating how to manage a disease, medications, and other treatments. Strategic planning creates a vision into the future that allows the planner to anticipate all kinds of possibilities and, most importantly, to be prepared for them.

Strategic planning powers your short-term and long-term goal setting. Once you have set goals, you need to plan the steps that will help you achieve those goals over time. For example, a strategic thinker who wants to own a home in five years' time might drive a used car and cut out luxuries, put a small amount of money every month into a mutual fund, and keep an eye on current mortgage percentages. In class, a strategic planner will think critically about the material presented, knowing that information is most useful later on if it is clearly understood.

Strategic planning helps you keep up with technology. As technology develops more and more quickly, some jobs become obsolete and others are created. It's possible to spend years in school training for a career area that will be drying up when you are ready to enter the work force. When you plan strategically, you can take a broader range of courses or choose a major and career that are expanding. This will make it more likely that your skills will be in demand when you graduate.

Effective critical thinking is essential to strategic planning. Here are some tips for becoming a strategic planner:

Develop a rational plan. What approach is most likely to achieve your goal? What steps will you need to take this year, or in 5, 10, or 20 years from now?

Anticipate all possible outcomes of your actions. What are the pros and cons?

Ask the question "how?" How do you achieve your goals? How do you learn effectively and remember what you learn? How do you develop a productive idea on the job? How do you distinguish yourself at school and at work?

Use human resources. Use information from talking to people who are where you want to be, professionally or personally. What caused them to get there? Ask them what they believe are the important steps to take, degrees to have, training to experience, and knowledge to gain.

In each thinking process, seen in Figure 5-4, use your imagination and creativity to come up with ideas, examples, causes, effects, and solutions. You have a capacity to be creative, whether you are aware of it or not. Open up your mind and awaken your creativity. It will enhance your critical thinking, make you a better science student, and make life more enjoyable.

HOW CAN YOU DEVELOP CREATIVITY IN SCIENCE?

Everyone is creative. Although the word "creative" may seem to refer primarily to artists, writers, musicians, and others who work in fields whose creative aspects are in the forefront, creativity comes in many other forms. It is the power to create anything, whether it is a solution, idea, approach, tangible product, work of art, system, program—anything at all. To help expand your concept of creativity, the list below offers examples of day-to-day creative thinking.

TERMS

Creativity
The ability to produce something new through imaginative skill.

Figure 5-4 The wheel of thinking.

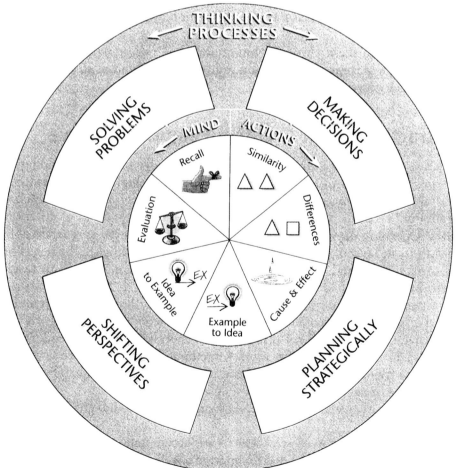

- Figuring out an alternative plan when your baby sitter unexpectedly cancels on you
- Planning how to coordinate your work and class schedules
- Talking through a problem with an instructor, and finding a way to understand each other

Creative innovations introduced by all kinds of people continually expand and change the world. Here are some that have had an impact:

- Susan B. Anthony and other women fought for and won the right for women to vote.
- Florence Nightingale promoted hand washing for health care workers and saved thousands of lives, even though the reason it helped, it eliminates bacteria, was not known.
- Henry Ford introduced the assembly line method of automobile construction, making cars cheap enough to be available to the average citizen.
- Rosa Parks refused to give up her seat on the bus to a white person, thus setting off a chain of events that gave rise to the civil rights movement.
- Alicia Diaz, director of the Center of Hispanic Policy, Research, and Development, developed corporate partnerships and internship programs that have become models for small, efficient government.

Even though these particular innovations had wide-ranging effects, the characteristics of these influential innovators can be found in all people who exercise their creative capabilities.

Characteristics of Creative People in Science

Creative people think in fresh new ways that improve our understanding of the world. Roger van Oech, an expert on creativity, highlights this kind of flexibility.[2] "I've found that the hallmark of creative people is their mental flexibility," he says. "Like race-car drivers who shift in and out of different gears depending on where they are on the course, creative people are able to shift in and out of different types of thinking depending on the needs of the situation at hand. . . . they're doggedly persistent in striving to reach their goals."

Creativity in science leads to many discoveries. Although many scientific discoveries occur accidentally, which is called *serendipity*, it takes creativity to recognize that the accidental discovery, or result, is valuable. Serendipity is common in science, it occurs when something is discovered while in the process of trying to discover something else. Serendipity is "the gift of finding valuable or agreeable things not sought for."[3]

Examples of creativity and serendipity in science abound and include:

- Velcro by George deMestral. Velcro was accidentally discovered by deMestral when he returned home from a walk and noticed a cocklebur stuck tightly to the fabric of his clothing. On close examination, he noticed how well the bur's hooks held onto the fabric and used that same system to make Velcro.

- Penicillin by Sir Alexander Fleming. In 1928, Fleming worked in a bacteriology laboratory preparing influenza cultures in petri dishes. He noticed a clear area in the dish where a piece of mold had fallen. He isolated the mold and found it belonged to the genus *Penicillium*. Out of the thousands of species of molds, this one happened to fall in Fleming's petri dish, but more importantly, he noticed its effect and followed up by asking why the bacteria culture was cleared away by this particular substance. Fleming's intelligence is what made the discovery. He is quoted as saying, "The story of penicillin has a certain romance in it and helps to illustrate the amount of chance, or fortune, or fate, or destiny, call it what you will, in anybody's career."[4]

- Prehistoric artifacts by Stephen Young. Young was a sociologist in Thailand who tripped on a root while walking down a road one day. As he lay face down in the dirt, he noticed pieces of pottery protruding from the earth; he guessed the pieces were very old because they were unglazed. An archaeologist later removed 18 tons of material and developed a new theory of the prehistory of Southeast Asia. Bronze jewelry and spear points were dated as early as 2000 B.C. Previously, it was thought that metallurgy wasn't in the area until as late as 500 B.C.

Discoveries in science are not dependent on accidents alone, creativity is needed to think of new ways to do things and to make sense of what you observe. Louis Pasteur, who made exceptional breakthroughs in microbiology and medicine, put it this way: "In the fields of observation, chance favors only the prepared mind." Yes, accidents do happen, and they can lead to major discoveries; but if you aren't ready, you won't know them when you see them. Unless the mind is thoroughly charged beforehand, the proverbial spark of genius, if it should manifest itself, probably will find nothing to ignite.[5]

Innovation comes to scientists in many forms. For the scientist who created the Mars landing team, *Pathfinder* and *Sojourner,* a scarcity of time and money were creative factors. Tony Spear, of the Pathfinder Project at the Jet Propulsion Laboratory, said: "Not only was landing on Mars tough, but we had to do it for a fixed cost and on a tight schedule."[6] Older tried-and-true methods were too expensive and were thrown out in favor of less-expensive innovations.

Another form of creativity is in asking the right questions. The 1998 Columbus scholar Marek Elbaum, a scientist at Electro-Optical Sciences, was interested in how he could help cancer specialists detect melanoma, one of the deadliest forms of cancer. He asked the following question and produced a creative answer:

We went to oncologists and we asked, "What is difficult for you?" And they said, "If you could help us discriminate early melanoma from benign pigmented lesions, this would be of great help." And we said with great arrogance, "Yes, we can do such things."[7]

The next question Elbaum and his colleagues asked was if the malignant lesions looked different than benign ones under different light frequencies.

This turned out to be the case, and they created an imaging device that detects melanoma in less than a minute.

Creating science out of everyday observations is what the web-based Beachcombers group and the Oceanography Society are doing by studying and recording the flotsam that washes up on beaches around the world. The idea behind this documentation is to discover social and economic clues about global shipping and how pollution travels, by harnessing members of the Beachcombers' natural enthusiasm for observation of vast stretches of the world's coastlines. The Web site, "Beachcombers' Alert!" provides potential data collectors instructions for joining the search.

Overall, your chance of making a miracle discovery in science is minimal, but you can significantly add to the body of knowledge that builds toward such discoveries. Nobel laureate Paul Flory, upon receiving the Priestley Medal, the American Chemical Society's highest award, said:

> Significant inventions are not mere accidents . . . happenstance usually plays a part, to be sure, but there is much more to invention than the popular notion of a bolt out of the blue. Knowledge in depth and breadth are virtual prerequisites.[8]

Brainstorming Toward a Creative Answer

As discussed earlier in the chapter, you are brainstorming when you approach a problem by letting your mind free-associate and come up with as many possible ideas, examples, or solutions as you can, without immediately evaluating them as good or bad. Brainstorming is also referred to as divergent thinking—you start with the issue or problem and then let your mind diverge, or go in as many different directions as it wants, in search of ideas or solutions. You can use brainstorming for problem solving, decision making, preparing to write an essay, or any time you want to open your mind to new possibilities. Here are some rules for successful brainstorming.[9]

Don't evaluate or criticize an idea right away. Write down your ideas so that you remember them. Evaluate later, after you have had a chance to think about them. Try to avoid criticizing other people's ideas as well. Students often become stifled when their ideas are evaluated during brainstorming. Notice your tendency to say that things are right or wrong, black or white, and instead let things be gray—unknown.

Focus on quantity; don't worry about quality until later. Try to generate as many ideas or examples as you can. The more thoughts you generate, the better the chance that one may be useful. Brainstorming works well in groups. Group members can become inspired by, and make creative use of, one another's ideas.

Let yourself consider wild and wacky ideas. Trust yourself to fall off the edge of tradition when you explore your creativity. Sometimes the craziest ideas end up being the most productive, positive, workable solutions around.

Remember, creativity can be developed if you have the desire and patience. Be gentle with yourself in the process. Most people are harsher with themselves and their ideas than is necessary. Your creative expression will become more free with practice.

Creativity and Critical Thinking

Critical thinking and creativity work hand in hand. Critical thinking is inherently creative, because it requires you to take the information you are given and come up with original ideas or solutions to problems. For example, you can brainstorm to generate possible causes of a certain effect. If the effect you were examining was fatigue in afternoon classes, you might come up with possible causes such as lack of sleep, too much morning caffeine, a diet heavy in carbohydrates, a natural tendency toward low energy at that time, or an instructor who doesn't inspire you. Through your consideration of causes and solutions, you have been thinking both creatively and critically.

Creative thinkers and critical thinkers have similar characteristics—both consider new perspectives, ask questions, don't hesitate to question accepted assumptions and traditions, and persist in the search for answers. Only through thinking critically and creatively can you freely question, brainstorm, and evaluate in order to come up with the most fitting ideas, solutions, decisions, arguments, and plans.

You use critical-thinking mind actions throughout everything you do in school and in your daily life.

> "The world of reality has its limits. The world of imagination is boundless."
>
> JEAN-JACQUES ROUSSEAU

Κρινειν

The word "critical" is derived from the Greek word *krinein*, which means to separate in order to choose or select. To be a mindful, aware critical thinker, you need to be able to separate, evaluate, and select ideas, facts, and thoughts.

Think of this concept as you apply critical thinking to your reading, writing, and interaction with others. Be aware of the information you take in and of your thoughts, and be selective as you process them. Critical thinking gives you the power to make sense of life by deliberately selecting how to respond to the information, people, and events that you encounter.

Chapter 5   Applications

Name _____ Date _____

KEY INTO YOUR LIFE
Opportunities to Apply What You Learn

 Brainstorming on the Idea Wheel

Your creative mind can solve problems when you least expect it. Many people report having sudden ideas while exercising, driving, showering, upon waking, or even when dreaming. When the pressure is off, the mind is often more free to roam through uncharted territory and bring back treasures.

To make the most of this "mind-float," grab ideas right when they surface. If you don't, they roll back into your subconscious as if on a wheel. Since you never know how big the wheel is, you can't be sure when that particular idea will roll to the top again. That's one of the reasons why writers carry notebooks—they need to grab thoughts when they come to the top of the wheel.

On the blank supplied below, name a problem, large or small, to which you haven't yet found a satisfactory solution. Do a brainstorm without the time limit. Be on the lookout for ideas, causes, effects, solutions, or similar problems coming to the top of your wheel. The minute it happens, grab this book and write your idea(s) next to the problem. Take a look at your ideas later, and see how your creative mind may have pointed you toward some original and workable solutions. You may want to keep a book by your bed to catch ideas that pop up before, during, or after sleep.

Problem:

Ideas:

_____

_____

_____

_____

_____

_____

_____

5.2 Brainstorming in a Group

In a group, brainstorm some answers to the following questions. Remember, in brainstorming there are no right or wrong answers. This is a time to be creative and let the ideas flow.

THEMES IN THE STUDY OF LIFE	QUESTIONS
Life is organized on many structural levels.	What are the structural levels?
Cells are the organism's basic unit of structure and function.	What purpose do cells serve in the function of organisms and their structure?
Continuity of life is based on inheritable information—DNA.	What are the structure and function of DNA in the inheritance of traits?
Study of organisms enriches the study of life.	What differences are there between an organism as a whole functioning unit and its structures?
Structure and function correlate at all levels of biological organization.	How does anatomy determine function?
Organisms are open systems that interact continuously with their environments.	How does the environment affect the function or structure of an organism?
Diversity and unity are the dual faces of life on earth.	In what ways do organisms differ and in which ways are they alike?
Evolution is the core theme of biology.	What are the hallmarks of evolution, and how does it unify, or connect, life on earth?
Science is a process of inquiry that often involves hypothetical and deductive thinking.	What is an example of deductive inquiry?

Source: Adapted from Campbell, N. A. (1997) *Biology* (4th ed.). Menlo Park: Benjamin Cummings, pages 2–24.

 ## 5.3 Essential Questioning

Choose one of the themes from Exercise 2. Considering this theme as a statement of a biological science principle, devise as many questions as you can that would help you clarify, validate, support, or dispute the principle.

5.4 Problem Solving

Choose two of your questions from Exercise 3 and answer the following:

A. What further information would be helpful to work toward an answer?

Question 1: _____

Question 2: _____

B. Where could you go for this information?

Question 1: _____

Question 2: _____

C. How you could go about testing your questions?

Question 1: _____

Question 2: _____

5.5 Biases

Consider the two questions you chose in Exercise 4 and the theme you chose for Exercise 3. Make a list of what biases may have affected your choices. (Try to think of at least three.)

_____

_____

_____

_____

_____

KEY TO SELF-EXPRESSION
Discovery Through Journal Writing

To record your thoughts, use a separate journal or the lined page at the end of the chapter.

Strategic Planning

Discuss your abilities and limitations in how you set and plan your short-term and long-term goals. Do you tend to plan ahead of time? Why or why not? What do you like and dislike about strategically planning ahead? Do you have a hard time seeing beyond the present, or do you like to predict what will happen in the future? Discuss a long-term goal in terms of what you want in one year, five years, ten years, and twenty years. How do you plan to accomplish this goal? What steps will you take? What do you want to achieve?

Journal

Name _____ Date _____

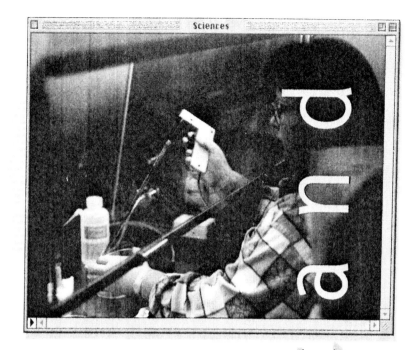

Reading and Studying

Maximizing Written Resources

Society revolves around the written word. As the *Condition of Education 1996* states, "In recent years, literacy has been viewed as one of the fundamental tools necessary for successful economic performance in industrialized societies. Literacy is no longer defined merely as a basic threshold of reading ability, but rather as the ability to understand and use printed information in daily activities, at home, at work, and in the community."[1]

If you read thoroughly and understand what you read and if you achieve your study goals, you can improve your capacity to learn and understand. In this chapter you will learn how you can overcome barriers to successful reading and benefit from defining a purpose every time you read. You will explore the PQ3R study technique and see how critical reading can help you maximize your understanding of any kind of science text. Finally, the chapter will provide an overview of your library's resources.

In this chapter, you will explore answers to the following questions:

- What are some challenges in science readings?
- What kind of reading will you do in the sciences?
- Why define your purpose for reading?
- How can PQ3R help you study reading materials?
- How can you read critically?
- What resources does your library offer?

WHAT ARE SOME CHALLENGES IN SCIENCE READINGS?

Whatever your skill level, you will encounter challenges that make reading more difficult, such as an excess of reading assignments, difficult texts, distractions, a lack of speed and comprehension, and insufficient vocabulary. Following are some ideas about how to meet these challenges. Note that if you have a reading disability, if English is not your primary language, or if you have limited reading skills, you may need additional support and guidance. Most colleges provide services for students through a reading center or tutoring program. Take the initiative to seek help if you need it. Many accomplished learners have benefited from help in specific areas.

Dealing With Reading Overload

Reading overload is part of almost every college experience. On a typical day, you may be faced with reading assignments that look like this:

- An entire textbook chapter on the causes of the Civil War (American history).
- An original research study on the stages of sleep (physiology).
- Pages 1–50 in a human biology textbook (biological science).

Reading all this and more leaves little time for anything else unless you read selectively and skillfully. You can't control your reading load. You can, however, improve your reading skills. The material in this chapter will present techniques that can help you read and study as efficiently as you possibly can, while still having time left over for other things.

Working Through Difficult Science Texts

While many science textbooks are useful teaching tools, some can be poorly written and organized. Students using texts that aren't well written may blame themselves for the difficulty they're experiencing. Because texts are often written with the purpose of challenging the intellect, even well-written and organized texts may be difficult and dense to read. Generally, the further you advance in your education, the more complex your required reading is likely to be. For example, your physics professor may assign a chapter on thermodynamics. You may feel at times as though you are reading a foreign language as you encounter new concepts, words, and terms.

Assignments can also be difficult when the required reading is from primary sources rather than from texts. *Primary sources* are original documents rather than another writer's interpretation of these documents. They include:

- Historical documents
- Scientific studies, including lab reports and accounts of experiments
- Research journal articles

The academic writing found in research journal articles and scientific studies is different from other kinds of writing. Some academic writers assume that readers understand sophisticated concepts. They may not define basic terms, provide background information, or supply a wealth of examples to support their ideas. As a result, concepts may be difficult to understand.

Making your way through poorly written or difficult reading material is hard work that can be accomplished through focus, motivation, commitment, and skill. The following strategies may help.

Approach your reading assignments head-on. Be careful not to prejudge them as impossible or boring before you even start to read.

Accept the fact that some texts may require some extra work and concentration. Set a goal to make your way through the material and learn, whatever it takes.

When a primary source discusses difficult concepts that it does not explain, put in some extra work to define such concepts on your own. Ask your instructor or other students for help. Consult reference materials in that particular subject area, other class materials, dictionaries, and encyclopedias. You may want to make this process more convenient by creating your own minilibrary at home. Collect reference materials that you use often, such as a dictionary, a thesaurus, a writer's style handbook, and maybe an atlas or computer manual. You may also benefit from owning reference materials in your particular areas of study. "If you find yourself going to the library to look up the same reference again and again, consider purchasing that book for your personal or office library," advises library expert Sherwood Harris.[2]

Look for order and meaning in seemingly chaotic reading materials. The information you will find in this chapter on the PQ3R reading technique and on critical reading will help you discover patterns and achieve a greater depth of understanding. Finding order within chaos is an important skill, not just in the mastery of reading, but also in life. This skill can give you power by helping you "read" (think through) work dilemmas, personal problems, and educational situations.

"No barrier of the senses shuts me out from the sweet, gracious discourse of my book friends. They talk to me without embarrassment or awkwardness."
HELEN KELLER

Managing Distractions

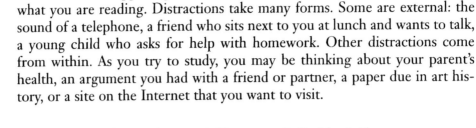

With so much happening around you, it's often hard to keep your mind on what you are reading. Distractions take many forms. Some are external: the sound of a telephone, a friend who sits next to you at lunch and wants to talk, a young child who asks for help with homework. Other distractions come from within. As you try to study, you may be thinking about your parent's health, an argument you had with a friend or partner, a paper due in art history, or a site on the Internet that you want to visit.

Identify the Distraction and Choose a Suitable Action

Pinpoint what's distracting you before you decide what kind of action to take. If the distraction is *external* and *out of your control*, such as construction outside your building or a noisy group in the library, try to move away from it. If the distraction is *external* but *within your control*, such as the television, telephone, or children, take action. For example, if the television or phone is a problem, turn off the TV or unplug the phone for an hour.

If the distraction is *internal*, there are a few strategies to try that may help you clear your mind. You may want to take a break from your studying and tend to one of the issues that you are worrying about. Physical exercise may relax you and bring back your ability to focus. For some people, studying while listening to music helps to quiet a busy mind. For others, silence may do the trick. If you need silence to read or study and cannot find a truly quiet environment, consider purchasing sound-muffling headphones or even earplugs.

Find the Best Place and Time to Read

Any reader needs focus and discipline in order to concentrate on the material. Finding a place and time that minimize outside distractions will help you achieve that focus. Here are some suggestions:

Read alone unless you are working with other readers. Family members, friends, or others who are not in study mode may interrupt your concentration. If you prefer to read alone, establish a relatively interruption-proof place and time, such as an out-of-the-way spot at the library or an after-class hour in an empty classroom. If you study at home and live with other people, you may want to place a "Quiet" sign on the door. Some students benefit from reading with one or more other students. If this helps you, plan to schedule a group reading meeting where you read sections of the assigned material and then break to discuss them.

Find a comfortable location. Many students study in the library on a hard-backed chair. Others prefer a library easy chair, a chair in their room, or even the floor. The spot you choose should be comfortable enough for hours of reading, but not so comfortable that you fall asleep. Also, make sure that you have adequate lighting and aren't too hot or too cold.

Choose a regular reading place and time. Choose a spot or two you like and return to them often. Also, choose a time when your mind is alert and focused. Some students prefer to read just before or after the class for which

REAL WORLD PERSPECTIVE

How can I cope with a learning disability?

Clacy Albert, Washington State University at Pullman, Communications Major

All my life I've felt different. I just couldn't seem to learn the way other kids did. I felt stupid and afraid that other people would think I couldn't do anything right. I wouldn't raise my hand in class because I was afraid of being laughed at. I wouldn't volunteer for games because I was afraid I'd let my team down. Study groups were impossible for me. I didn't want anyone to know that I was different. Because of this, my self-esteem really suffered. I became very quiet.

It wasn't until I was a sophomore in high school that a teacher recognized something was wrong with the way I learned. It was my math teacher who saw that I couldn't recognize certain patterns. I would see things in reverse or not be able to recognize a pattern at all. He sat down with my parents and helped them understand something was wrong. Unfortunately, the school I attended didn't have any testing for learning disabilities, so I let it go until I was in college. When I enrolled at WSU, they told us about the learning disability resource center. My mom suggested I finally get the testing I needed. I'm glad I did, because now I know that I have dyslexia and need special assistance to handle my studies. I wish there was mandatory testing for this disability in grade school. If there had been, I wouldn't have suffered so deeply all these years. What suggestions do you have for helping me cope with this disability?

Edith Hall, Senior Sales Representative—Prentice Hall

I have a different disability but one that causes similar problems. I have attention deficit hyperactivity disorder (ADHD), and the fact that it was undiagnosed and untreated for many years has caused lots of problems in my life. It wasn't until I was six years out of college that I was diagnosed ADHD. And the great thing about it is I don't feel crazy anymore. Now I know why I can't sit still for long periods and why I can't complete large and/or long projects like other non-ADHD people can.

I think acknowledging that I had a disorder and then accepting it were the biggest steps to coping and living with this disorder. The other thing I have done is to get educated. I have read almost anything I can get my hands on. I am also involved in a support group. Having other people I can talk with about how my brain affects my behavior and my life truly is one of the best coping strategies I know.

Having a disability or disorder is not a bad thing. Ennis Cosby, slain son of comedian Bill Cosby, said of his dyslexia, "The day I found out I had dyslexia was the best day of my life." Finding out he had dyslexia relieved him of the belief that he was dumb or stupid or slow. For me, like Ennis Cosby, finding out I had ADHD was a great day in my life because I now had tools and help to be different . . . and I no longer felt alone.

the reading is assigned. Eventually, you will associate preferred places and times with focused reading.

If it helps you concentrate, listen to soothing background music. The right music can drown out background noises and relax you. However, the wrong music can make it impossible to concentrate; for some people, silence is better. Experiment to learn what you prefer; if music helps, stick with the type that works best. A personal headset makes listening possible no matter where you are.

Turn off the television. For most people, reading and TV don't mix.

Building Comprehension and Speed

Most students lead busy lives, carrying heavy academic loads while perhaps working a job or even caring for a family. It's difficult to make time to study at all, let alone handle the enormous reading assignments for your different classes. Increasing your reading comprehension and speed will save you valuable time and effort.

Rapid reading won't do you any good if you can't remember the material or answer questions about it. However, reading too slowly can be equally inefficient because it often eats up valuable study time and gives your mind space to wander. Your goal is to read for maximum speed *and* comprehension. Focus on comprehension first, because greater comprehension is the primary goal and also promotes greater speed.

Methods for Increasing Reading Comprehension

Following are some specific strategies for increasing your understanding of what you read:

Continually build your knowledge through reading and studying. More than any other factor, what you already know before you read a passage will determine your ability to understand and remember important ideas. Previous knowledge, including vocabulary, facts, and ideas, gives you a context for what you read.

TERMS
Context
Written or spoken knowledge that can help to illuminate the meaning of a word or passage.

Establish your purpose for reading. When you establish what you want to get out of your reading, you will be able to determine what level of understanding you need to reach and, therefore, on what you need to focus.

Remove the barriers of negative self-talk. Instead of telling yourself that you cannot understand, think positively. Tell yourself: *I can learn this material. I am a good reader.* And then, if you do need help, get it.

Think critically. Ask yourself questions. Do you understand the sentence, paragraph, or chapter you just read? Are ideas and supporting examples clear to you? Could you clearly explain what you just read to someone else?

Methods for Increasing Reading Speed

The following suggestions will help increase your reading speed.

- Try to read groups of words rather than single words.
- Avoid pointing your finger to guide your reading, since this will slow your pace.

- Try swinging your eyes from side to side as you read a passage, instead of stopping at various points to read individual words.
- When reading narrow columns, focus your eyes in the middle of the column and read down the page. With practice, you'll be able to read the entire column width.
- Avoid vocalization when reading.
- Avoid thinking each word to yourself as you read it, a practice known as *subvocalization*. Subvocalization is one of the primary causes of slow reading speed.

Vocalization The practice of speaking the words and/or moving your lips while reading.

Facing the challenges of reading is only the first step. The next important step is to look at the different kinds of reading you will encounter as a science major.

WHAT KIND OF READING WILL YOU DO IN THE SCIENCES?

Readings in science courses will differ from liberal arts courses in three ways.

1. Amount of new vocabulary you will need to learn. Any science you study will have new terms to describe phenomena unique to that discipline. For example, in biology words crop up like chloroplasts, isotonic, plasmolysis, and vestigial. In nursing and other professions, acronyms are the rule rather than the exception. A patient has an AMI (anterior myocardial infarction), requires q 1 ABGs (arterial blood gases monitored every hour), and is on an IABP (intra-aortic balloon pump) in preparation for a PTCA (percutaneous transluminal angioplasty) or a CABG (coronary artery bypass graft).

2. Content is generally not written in a narrative style. Therefore, following the flow of the text is more challenging. Science books, if written in typical technical writing style, can be dry and a drudgery to plow through. Fortunately, there are many technical and textbook writers who are able to present scientific information in lively and interesting ways. Hopefully, these writers will be the ones you are required to read. If not, just take extra time for your reading and ask for help interpreting the "foreign" language you are learning.

3. Content is concentrated and text is full of information, diagrams, and formulas. Science texts are concentrated, usually lacking any story, or as described above, lacking easy narrative flow. In addition, you are required to read and understand symbols, math formulas and equations, diagrams of models, and graphs, which are used to explain the material. You will be memorizing new terms and models in order to learn more complex concepts. Graphics, that is, not written text, are often used to represent ideas.

Along with your readings from textbooks, you will read from research journals and possibly other science-oriented publications like *Scientific American* or *Natural History*. Books hold up-to-date information but not the most current, cutting edge information. From the time an author submits her manuscript for a textbook, it can take a year or more to be printed. For basic, or beginning, science courses, such as geology, physiology, or chemistry, the age of a textbook matters less than for courses dealing with information on technology or breakthrough discoveries, such as new cancer treatments.

The most up-to-date information is found in journals that publish research findings. Journals are designed to disseminate new and relevant information to the scientific community of a particular field. In nursing there is the *Journal of Advanced Nursing Science* and *Image;* in medicine, *JAMA* and the *New England Journal of Medicine;* in life sciences there is *Nature*. These are all examples of discipline-specific publications that publish research articles.

Journals serve the function of letting other researchers, or would-be researchers, see what their colleagues are doing. Repetition of studies is an important step in validating results before they are put into practice or applied to further study. Many research studies are designed to replicate research found in journals.

Another purpose of journals is to critically review the research being submitted for publication. Journals have reviewers who are asked to read and comment on an article before it is published. This review process helps ensure that published research is of high quality. However, not all research that is published is of the highest quality, and this is one very good reason why you will be learning to review research articles yourself. You have to understand the concepts of research and research methodology before you continue in a science career, because once in the field you must be able to critically review the research you read before you put it into practice.

Now, you understand better the challenges and types of reading in science. The final important step is to examine why you are reading any given piece of material.

WHY DEFINE YOUR PURPOSE FOR READING?

As with all other aspects of your education, asking important questions will enable you to make the most of your efforts. When you define your purpose, you ask yourself *why* you are reading a particular piece of material. One way to do this is by completing this sentence: "In reading this material, I intend to define/learn/answer/achieve . . ." With a clear purpose in mind, you can decide how much time and what kind of effort to expend on various reading assignments. Nearly 375 years ago, Francis Bacon, the English philosopher, recognized that

> Some books are to be tasted, others to be swallowed, and some few to be chewed and digested; that is, some books are to be read only in parts, others to be read but not curiously; and some few to be read wholly, and with diligence and attention.

Achieving your reading purpose requires adapting to different types of reading materials. Being a flexible reader—adjusting your reading strategies and pace—will help you to adapt successfully.

Purpose Determines Reading Strategy

With purpose comes direction; with direction comes a strategy for reading. Following are four reading purposes, examined briefly. You may have one or more for each piece of reading material you approach.

Purpose 1: Read to evaluate critically. Critical evaluation involves approaching the material with an open mind, examining causes and effects, evaluating ideas, and asking questions that test the strength of the writer's argument and that try to identify assumptions. Critical reading is essential for you to demonstrate an understanding of material that goes beyond basic recall of information. You will read more about critical reading later in the chapter.

Purpose 2: Read for comprehension. Much of the studying you do involves reading for the purpose of comprehending the material. The two main components of comprehension are *general principles* and *specific facts/examples*. These components depend on one another. Facts and examples help to explain or support ideas, and ideas provide a framework that helps the reader to remember facts and examples.

> *General Principles.* General principles reading is rapid reading that seeks an overview of the material. You may skip entire sections as you focus on headings, subheadings, and summary statements in search of general principles.
>
> *Specific Facts/Examples.* At times, readers may focus on locating specific pieces of information—for example, the sequence of geologic time periods. Often, a reader may search for examples that support or explain more general principles—for example, the structure of a plant cell. Because you know exactly what you are looking for, you can skim the material at a rapid rate. Reading your texts for specific information may help before taking a test.

Purpose 3: Read for practical application. A third purpose for reading is to gather usable information that you can apply toward a specific goal. When you read a computer software manual, an instruction sheet for assembling a bookshelf, or a cookbook recipe, your goal is to learn how to do something. Reading and action usually go hand in hand.

Purpose 4: Read for pleasure. Some materials you read for entertainment, such as *Rolling Stone* magazine or the latest John Grisham courtroom thriller. Entertaining reading may also go beyond materials that seem obviously designed to entertain. Whereas some people may read a Jane Austen novel for comprehension, as in a class assignment, others may read Austen's books for pleasure.

Purpose Determines Pace

George M. Usova, senior education specialist and graduate professor at the Johns Hopkins University, explains: "Good readers are flexible readers. They read at a variety of rates and adapt them to the reading *purpose* at hand, the *difficulty* of the material, and their *familiarity* with the subject area."[3] As Table 6-1 shows, good readers link the pace of reading to their reading purpose.

So far, this chapter has focused on reading. Recognizing obstacles to effective reading and defining the various purposes for reading lay the groundwork for effective *studying*—the process of mastering the concepts and skills contained in your texts.

HOW CAN PQ3R HELP YOU STUDY READING MATERIALS?

When you study, you take ownership of the material you read. You learn it well enough to apply it to what you do. For example, by the time students studying to be computer hardware technicians complete their course work,

Table 6-1 Linking purpose to pace.

TYPE OF MATERIAL	READING PURPOSE	PACE
Academic readings ◆ Textbooks ◆ Original sources ◆ Articles from scholarly journals ◆ Online publications for academic readers ◆ Lab reports ◆ Required fiction	◆ Critical analysis ◆ Overall mastery ◆ Preparation for tests	◆ Slow, especially if the material is new and unfamiliar
Manuals ◆ Instructions ◆ Recipes	◆ Practical application	◆ Slow to medium
Journalism and nonfiction for the general reader ◆ Nonfiction books ◆ Newspapers ◆ Magazines ◆ Online publications for the general public	◆ Understanding of general ideas, key concepts, and specific facts for personal understanding and/or practical application	◆ Medium to fast
Nonrequired Fiction	◆ Understanding of general ideas, key concepts, and specific facts for enjoyment	◆ Variable, but tending toward the faster speeds

Source: Adapted from Nicholas Reid Schaffzin, *The Princeton Review Reading Smart.* New York: Random House, 1996, p. 15.

they should be able to assemble various machines and analyze hardware problems that lead to malfunctions.

Studying also gives you mastery over *concepts*. For example, a dental hygiene student learns the causes of gum disease, a biology student learns what happens during photosynthesis, and a health sciences student learns about health policy analysis.

This section will focus on a technique that will help you learn and study more effectively as you read your college textbooks.

Preview-Question-Read-Recite-Review (PQ3R)

PQ3R is a technique that will help you grasp ideas quickly, remember more, and review effectively and efficiently for tests. The symbols PQ3R stand for *preview, question, read, recite,* and *review*—all steps in the studying process. Developed more than 55 years ago by Francis Robinson, the technique is still being used today because it works.[4] It is particularly helpful for studying science texts.

Moving through the stages of PQ3R requires that you know how to skim and scan. Skimming involves rapid reading of various chapter elements, including introductions, conclusions, and summaries; the first and last lines of paragraphs; boldface or italicized terms; pictures, charts, and diagrams. In contrast, scanning involves the careful search for specific facts and examples. You will probably use scanning during the *review* phase of PQ3R when you need to locate and remind yourself of particular information. In a chemistry text, for example, you may scan for examples of how to apply a particular formula.

Skimming
Rapid, superficial reading of material that involves glancing through to determine central ideas and main elements.

Scanning
Reading material in an investigative way, searching for specific information.

Preview

The best way to ruin a whodunit novel is to flip through the pages to find out how everything turned out. However, when reading textbooks, previewing can help you learn and is encouraged. *Previewing* refers to the process of surveying, or pre-reading, a book before you actually study it. Most textbooks include devices that give students an overview of the text as a whole as well as of the contents of individual chapters. As you look at Figure 6-1, on the following page, think about how many of these devices you already use.

Question

Your next step is to examine the chapter headings and, on your own paper, write questions linked to those headings. These questions will focus your attention and increase your interest, helping you relate new ideas to what you already know and building your comprehension. You can take questions from the textbook or from your lecture notes, or come up with them on your own when you preview, based on what ideas you think are most important.

Here is how this technique works. In the box on page 149, the column on the left contains primary- and secondary-level headings from a chapter of *Biology*, a text by Neil A. Campbell. The column on the right rephrases these headings in question form.

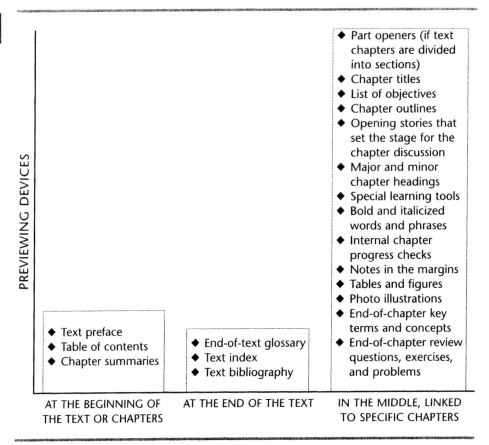

Figure 6-1
Text and chapter previewing devices.

There is no "correct" set of questions. Given the same headings, you would create your own particular set of questions. The more useful kinds of questions are ones that engage the critical-thinking mind actions and processes found in Chapter 5.

Read

Your questions give you a starting point for *reading*, the first R in PQ3R. Read the material with the purpose of answering each question you raised. Pay special attention to the first and last lines of every paragraph, which should tell you what the paragraph is about. As you read, record key words, phrases, and concepts in your notebook. Some students divide the notebook into two columns, writing questions on the left and answers on the right. This method, known as the Cornell note-taking system, is described in more detail in Chapter 7.

If you own the textbook, marking it up—in whatever ways you prefer—is a must. The notations that you make will help you to interact with the material and make sense of it. You may want to write notes in the margins, circle key ideas, or highlight key sections. Some people prefer to underline, although underlining adds more ink to the lines of text and may overwhelm your eye. Although writing in a textbook makes it difficult to sell it back to the bookstore, the increased depth of understanding you can gain is worth the investment.

THEMES IN THE STUDY OF LIFE	QUESTIONS
Life is organized on many structural levels.	What are the structural levels?
Cells are the organism's basic unit of structure and function.	What purpose do cells serve in the function of organisms and their structure?
Continuity of life is based on inheritable information—DNA.	What are the structure and function of DNA in the inheritance of traits?
Study of organisms enriches the study of life.	What differences are there between an organism as a whole functioning unit and its structures?
Structure and function correlate at all levels of biological organization.	How does anatomy determine function?
Organisms are open systems that interact continuously with their environments.	How does the environment affect the function or structure of an organism?
Diversity and unity are the dual faces of life on earth.	In what ways do organisms differ and in which ways are they alike?
Evolution is the core theme of biology.	What are the hallmarks of evolution, and how does it unify, or connect, life on earth?
Science is a process of inquiry that often involves hypothetical and deductive thinking.	What is an example of deductive inquiry?

Source: Adapted from Campbell, N. A. (1997). *Biology* (4th ed.). Menlo Park: Benjamin Cummings, pages 2–24.

Highlighting may help you pinpoint material to review before an exam. Here are some additional tips on highlighting:

Get in the habit of marking the text *after* you read the material. If you do it while you are reading, you may wind up marking less important passages.

Highlight key terms and concepts. Mark the examples that explain and support important ideas. You might try highlighting ideas in one color and examples in another.

Highlight figures and tables. They are especially important if they summarize text concepts.

Avoid overmarking. A phrase or two is enough in most paragraphs. Set off long passages with brackets rather than marking every line.

Write notes in the margins with a pen or pencil. Comments like "main point" and "important definition" will help you find key sections later on.

Be careful not to mistake highlighting for learning. You will not necessarily learn what you highlight unless you review it carefully. You may benefit from writing the important information you have highlighted into your lecture notes.

One final step in the reading phase is to divide your reading into digestible segments. Many students read from one topic heading to the next, then stop. Pace your reading so that you understand as you go. If you find you are losing the thread of the ideas you are reading, you may want to try smaller segments, or you may need to take a break and come back to it later.

Recite

Once you finish reading a topic, stop and answer the questions you raised about it in the Q stage of PQ3R. You may decide to recite each answer aloud, silently speak the answers to yourself, tell the answers to another person as though you were teaching him or her, or write your ideas and answers in brief notes. Writing is often the most effective way to solidify what you have read. Use whatever techniques best suit your learning style profile (see Chapter 3).

After you finish one section, move on to the next. Then repeat the question-read-recite cycle until you complete the entire chapter. If during this process you find yourself fumbling for thoughts, it means that you do not yet "own" the ideas. Reread the section that's giving you trouble until you master its contents. Understanding each section as you go is crucial because the material in one section often forms a foundation for the next.

Review

Review soon after you finish a chapter. Here are some techniques for reviewing.

- Skim and reread your notes. Then try summarizing them from memory.
- Answer the text's end-of-chapter review, discussion, and application questions.
- Quiz yourself, using the questions you raised in the Q stage. If you can't answer one of your own or one of the text's questions, go back and scan the material for answers.
- Review and summarize in writing the sections and phrases you have highlighted.
- Create a chapter outline in standard outline form or think-link form.
- Reread the preface, headings, tables, and summary.
- Recite important concepts to yourself, or record important information on a cassette tape and play it on your car's tape deck or your Walkman.
- Make flashcards that have an idea or word on one side and examples, a definition, or other related information on the other. Test yourself.
- Think critically: Break ideas down into examples, consider similar or different concepts, recall important terms, evaluate ideas, and explore causes and effects.
- Make think links that show how important concepts relate to one another.

Remember that you can ask your instructor if you need help clarifying your reading material. Your instructor is an important resource. Pinpoint the material you want to discuss, schedule a meeting with her during office hours, and come prepared with a list of questions. You may also want to ask what materials to focus on when you study for tests.

If possible, you should review both alone and with study groups. Reviewing in as many different ways as possible increases the likelihood of retention. Figure 6-2 shows some techniques that will help a study group maximize its time and efforts.

Repeating the review process renews and solidifies your knowledge. That is why it is important to set up regular review sessions—for example, once a week. As you review, remember that refreshing your knowledge is easier and faster than learning it the first time.

As you can see in Table 6-2 on p. 153, using PQ3R is part of being an active reader. Active reading involves the specific activities that help you retain what you learn.

How CAN YOU READ CRITICALLY?

Your textbooks will often contain features that highlight important ideas and help you determine questions to ask while reading. As you advance in your education, however, many reading assignments will not be so clearly marked, especially if they are primary sources such as research reports in journals. You will need critical-reading skills in order to select the important ideas, identify examples that support them, and ask questions about the text without the aid of any special features or tools.

Critical reading enables you to consider reading material carefully, developing a thorough understanding of it through evaluation and analysis. A critical reader is able to discern what in a piece of reading material is true or useful, such as when using material as a source for an essay. A critical reader can also compare one piece of material to another and evaluate which makes more sense, which proves its thesis more successfully, or which is more useful for the reader's purpose.

Critical reading is reading that transcends taking in and regurgitating material. You can read critically by using PQ3R to get a basic idea of the material, asking questions based on the critical-thinking mind actions, shifting your perspective, and seeking understanding.

Use PQ3R to "Taste" Reading Material

Sylvan Barnet and Hugo Bedau, authors of *Critical Thinking, Reading, and Writing—A Brief Guide to Argument*, suggest that the active reading of PQ3R will help you form an initial idea of what a piece of reading material is all about. Through previewing, skimming for ideas and examples, highlighting and writing comments and questions in the margins, and reviewing, you can develop a basic understanding of its central ideas and contents.[5]

Summarizing, part of the review process in PQ3R, is one of the best ways to develop an understanding of a piece of reading material. To construct a

152 CHAPTER 6 Reading and Studying

Figure 6-2   Study group techniques.

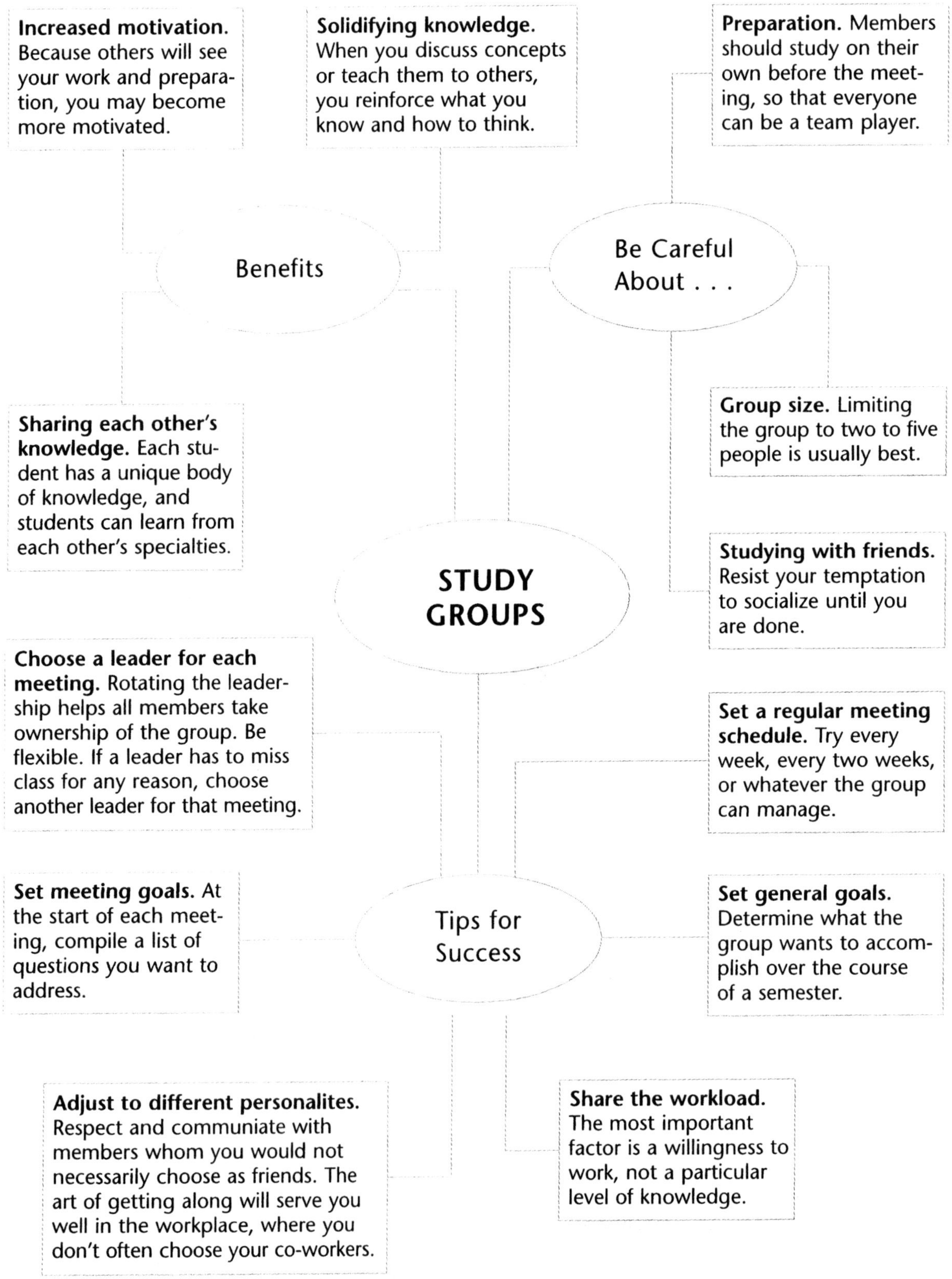

STUDY GROUPS

Benefits

Increased motivation. Because others will see your work and preparation, you may become more motivated.

Solidifying knowledge. When you discuss concepts or teach them to others, you reinforce what you know and how to think.

Sharing each other's knowledge. Each student has a unique body of knowledge, and students can learn from each other's specialties.

Be Careful About . . .

Preparation. Members should study on their own before the meeting, so that everyone can be a team player.

Group size. Limiting the group to two to five people is usually best.

Studying with friends. Resist your temptation to socialize until you are done.

Tips for Success

Choose a leader for each meeting. Rotating the leadership helps all members take ownership of the group. Be flexible. If a leader has to miss class for any reason, choose another leader for that meeting.

Set meeting goals. At the start of each meeting, compile a list of questions you want to address.

Adjust to different personalites. Respect and communiate with members whom you would not necessarily choose as friends. The art of getting along will serve you well in the workplace, where you don't often choose your co-workers.

Set a regular meeting schedule. Try every week, every two weeks, or whatever the group can manage.

Set general goals. Determine what the group wants to accomplish over the course of a semester.

Share the workload. The most important factor is a willingness to work, not a particular level of knowledge.

Use PQ3R to become an active reader

ACTIVE READERS TEND TO...

Divide material into manageable sections	Answer end-of-chapter questions and applications
Write questions	Create chapter outlines
Answer questions through focused note taking	Create think links that map concepts in a logical way
Highlight key concepts	Make flash cards and study them
Recite, verbally and in writing, the answers to questions	Recite what they learned into a tape recorder and play the tape back
Focus on main ideas found in paragraphs, sections, and chapters	Rewrite and summarize notes and highlighted materials from memory
Recognize summary and support devices	Explain what they read to a family member or friend
Analyze tables, figures, and photos	Form a study group

summary, focus on the central ideas of the piece and the main examples that support those ideas. A summary does not contain any of your own ideas or your evaluation of the material. It simply condenses the material, making it easier for you to focus on the structure of the piece and its central ideas when you go back to read more critically. At that point, you can begin to evaluate the piece and introduce your own ideas. Using the mind actions will help you.

TERMS

Summary
A concise restatement of the material, in your own words, that covers the main points.

Ask Questions Based on the Mind Actions

The essence of critical reading, as with critical thinking, is asking questions. Instead of simply accepting what you read, seek a more thorough understanding by questioning the material as you go along. Using the mind actions of the Thinktrix to formulate your questions will help you understand the material.

What parts of the material you focus on will depend on your purpose for reading. For example, if you are writing a paper on the effects of radiation on man-in-the-moon marigolds, you might spend your time focusing on how certain causes fit your thesis. If you are comparing two pieces of writing that contain opposing arguments, you may focus on picking out their central ideas and evaluating how well the writers use examples to support these ideas.

You can question any of the following components of reading material:

- The central idea of the entire piece
- A particular idea or statement
- The examples that support an idea or statement
- The proof of a fact
- The definition of a concept

Following are some ways to critically question your reading material, based on the mind actions. Apply them to any component you want to question by substituting the component for the words "it" and "this."

Similarity: What does this remind me of, or how is it similar to something else I know?

Difference: What different conclusions are possible?
How is this different from my experience?

Cause and Effect: Why did this happen, or what caused this?
What are the effects or consequences of this?
What effect does the author want to have, or what is the purpose of this material?
What effects support a stated cause?

Example to Principle: How would I classify this, or what is the best principle to fit this example(s)?
How would I summarize this, or what are the key principles?
What is the thesis or central idea?

Principle to Example: What evidence supports this, or what examples fit this principle?

Evaluation: How would I evaluate this? Is it valid or pertinent?
Does this example support my thesis or central idea?

Shift Your Perspective

Your understanding of perspective will help you understand that many reading materials are written from a particular perspective. Perspective often has a strong effect on how the material is presented. For example, if a recording artist and a music censorship advocate were each to write a piece about a controversial song created by that artist, their different perspectives would result in two very different pieces of writing.

To analyze perspective, ask questions like the following:

What perspective is guiding this? What are the underlying ideas that influence this material?

Who wrote this, and what may be the author's perspective? For example, a piece on a new drug written by an employee of the drug manufacturer may differ from a doctor's evaluation of the drug.

What does the title of the material tell me about its perspective? For example, a piece entitled "New Therapies for Diabetes" may be more informational, and "What's Wrong with Insulin Injections" may intend to be persuasive.

How does the material's source affect its perspective? For example, an article on health management organizations (HMOs) published in an HMO newsletter may be more favorable and one-sided than one published in *The New York Times*.

Seek Understanding

Reading critically allows you to investigate what you read so that you can reach the highest possible level of understanding. Think of your reading process as an archaeological dig. The first step is to excavate a site and uncover the artifacts. In reading, that corresponds to your initial preview and reading of the material. As important as the excavation is, the process would be incomplete if you stopped there and just took home a bunch of items covered in dirt. The second half of the process is to investigate each item, evaluate what all of those items mean, and to derive new knowledge and ideas from what you discover. Critical reading allows you to complete that crucial second half of the process.

As you work through all of the different requirements of critical reading, remember that critical reading takes *time* and *focus*. Finding a time, place, and purpose for reading, covered earlier in this chapter, is crucial to successful critical reading. Give yourself a chance to gain as much as possible from what you read.

No matter where or how you prefer to study, your school's library (or libraries) can provide many useful services to help you make the most of classes, reading, studying, and assignments.

HAT RESOURCES DOES YOUR LIBRARY OFFER?

Your library can help you search for all kinds of information. First, learn about your library, its resources, and its layout. While some schools have only one library, other schools have a library network that includes one or more central libraries and smaller, specialized libraries that focus on specific academic areas. Take advantage of library tours, orientation, training sessions, and descriptive pamphlets. Spend time walking around the library on your own. If you still have questions, ask a reference librarian. A simple question can save hours of searching. The following sections will help you understand how your library operates.

"With one day's reading a man may have the key in his hands."

EZRA POUND

General Reference Works

As a science student, you will spend time in the library getting to know, and use, the reference section. Getting to know the reference librarian, the library layout, and resources will save you time later on.

General reference works give you an overview and lead you to more specific information. These works cover topics in a broad, nondetailed way. General reference guides are found in the front of most libraries and are often available on CD-ROM, a compact disk that contains millions of words and images. You access this information by inserting the disk into a specially designed computer. Among the works that fall into this category are:

- encyclopedias—for example, the multivolume *Encyclopedia Britannica*
- almanacs—*The World Almanac and Book of Facts*
- dictionaries—*Webster's New World College Dictionary*
- biographical reference works—*Webster's Biographical Dictionary*
- bibliographies—*Books in Print*

Specialized Reference Works

Look at *specialized reference works* to find more specific facts. Specialized reference works include encyclopedias and dictionaries that focus on a narrow field. The short summaries you will find there focus on critical ideas. Bibliographies that accompany the articles point you to the names and works of recognized experts. Examples of specialized references include the *International Encyclopedia of Film*, the *Encyclopedia of Computer Science and Technology*, and the *Dictionary of Education*.

Library Book Catalog

Found near the front of the library, the *book catalog* lists every book the library owns. The listings usually appear in three separate categories: authors' names, book titles, and subjects. Not too long ago, most libraries stored their book catalogs on index-sized cards in hundreds of small drawers. Today, many libraries have replaced these cards with computers. Using a terminal that has access to the library's computer records, you can search by specific author, title, and subject.

The computerized catalog in your college library is probably connected to the holdings of other college and university libraries. This gives you an on-line search capacity, which means that if you don't find the book you want in your local library, you can track it down in another library and request it through an interlibrary loan. *Interlibrary loan* is a system used by many colleges to allow students to borrow materials from a library other than the one at their school. Students request materials through their own library, where the materials are eventually delivered by the outside library. When you are in a rush, keep in mind that interlibrary loans may take a substantial amount of time.

Periodical Indexes

Periodicals are magazines, journals, and newspapers that are published on a regular basis throughout the year. Examples include *Nature, Computer Science*, and *Science*. Many libraries display periodicals that are a year or two old and convert older copies to microfilm or microfiche (photocopies of materials reduced greatly in size and printed on film readable in a special reading machine—*microfilm* is a strip of film, and *microfiche* refers to individual leaves of film). Finding articles in publications involves a search of periodical indexes. The most widely used general index is the *Reader's Guide to Periodical Literature*, which is available on CD-ROM and in book form. The *Reader's Guide* indexes articles in more than 100 general-interest magazines and journals.

Electronic Research

You will also find complete source material through a variety of electronic sources, including the Internet, on-line services, and CD-ROM. Here is a sampling of the kind of information you will find:

- complete articles from thousands of journals and magazines
- complete articles from newspapers around the world
- government data on topics as varied as agriculture, transportation, and labor
- business documents, including corporate annual reports

Your library is probably connected to the *Internet*, a worldwide computer network that links government, university, research, and business computers along the Information Superhighway. Tapping into the *World Wide Web*—a tool for searching the huge libraries of information stored on the Internet—gives you access to billions of written words and graphic images. The Internet is so vast that you may need other publications to help you explore it. Seek out tools to aid you on your journeys along the Information Superhighway. If your college has its own Internet home page, start by spending some time browsing through it.

Although most libraries do not charge a fee to access the Internet, they may charge when you connect to commercial on-line services, including Nexis, CompuServe, and Prodigy. Ask your librarian about all fees and restrictions. Libraries also have electronic databases on CD-ROM. A database is a collection of data—or, in the case of most libraries, a list of related resources that all focus on one specific subject area—arranged so that you can search through it and retrieve specific items easily. For example, the DIALOG Information System includes hundreds of small databases in specialized areas. CD-ROM databases are generally smaller than on-line databases and are updated less frequently. However, there is never a user's fee.

читать

This word may look completely unfamiliar to you, but anyone who can read the Russian language and alphabet will know that it means "read." People who read languages that use different kinds of characters, such as Russian, Japanese, or Greek, learn to process those characters as easily as you process the letters of your native alphabet. Your mind learns to process individually each letter or character you see. This ability enables you to move to the next level of understanding—making sense of those letters or characters when they are grouped to form words, phrases, and sentences.

Think of this concept when you read. Remember that your mind is an incredible tool, processing unmeasurable amounts of information so that you can understand the concepts on the page. Give it the best opportunity to succeed by reading as often as you can and by focusing on all of the elements that help you read to the best of your ability.

Chapter 6 Applications

Name _____ Date _____

KEY INTO YOUR LIFE
Opportunities to Apply What You Learn

 6.1 Studying a Text Page

The following excerpt is from Evaluation and Exercise Prescription, Fardy and Yanowitz, *Cardiac Rehabilitation, Adult Fitness, and Exercise Testing*, 3rd ed. Read the material using the study techniques in this chapter, and complete the questions that follow.

EXERCISE PRESCRIPTION

THE EXERCISE PRESCRIPTION represents the carefully regulated dosage of physical activity of a long-term training program. The dosage consists of a coalescence of intensity, duration, and frequency of effort that is undertaken in an exercise mode to achieve specific program objectives. The prescription should be developed in a manner similar to that of prescribing medication, that is, administered in specified amounts based on individual needs. When designed for cardiac rehabilitation and adult fitness, the exercise modes are usually selected for the purpose of enhancing cardiovascular function and lessening the risk of coronary heart disease. Other training objectives such as strength and flexibility have very different prescriptions and are addressed briefly in this chapter, although the reader is referred elsewhere for more in-depth information. The purposes of this chapter are to present the rationale for an exercise prescription that promotes cardiovascular function; to present the physiologic basis and design of the prescription, the factors that affect the prescription; and to apply the prescription formula to an exercise training program.

Physiologic Basis of the Exercise Prescription

The physiologic basis of the exercise prescription is the overload principle and the relationship between training stimuli (dosage) and adaptation (response).

Overload Principle

Overload by definition means that the training stimulus must surpass normal daily physical exertion to be beneficial. The training stimulus, however, should not provoke undue fatigue, musculoskeletal strain, or mental or emotional burnout. Optimal benefit necessitates regular updating of the overload threshold.

Dose Response

Adaptation is related to the amount of physical exertion, although the relationship is not consistently linear. Dose-response curves depicted in Figure A.1 represent a relationship illustrating that adaptation does not occur until some minimal effort is expended, that is, overload. The curves do not represent physiologic measures, but rather represent a conceptual comparison of effort versus gain under different circumstances.

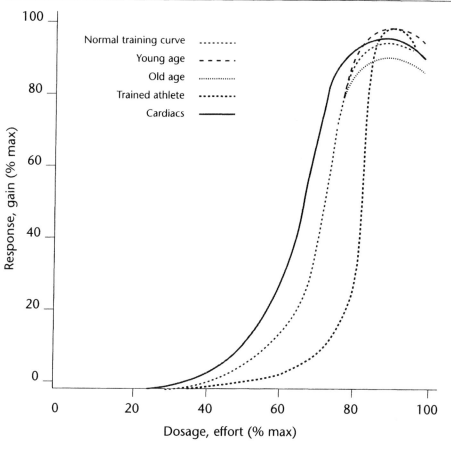

FIGURE A-1
Improvement anticipated from effort expended.

Improvement anticipated from effort expended: A conceptual diagram illustrating training curves in various populations. Note that the trained athlete is required to exercise at a greater intensity to make gains similar to those of normal persons of cardiac patients. Also note that age at onset of training affects the maximal physiologic gain. (Redrawn from Fardy PS. "Train for aerobic power." In: Burke E.J., ed., *Toward an Understanding of Human Performance*. Ithaca, NY: Movement Publications, 1977).

Training adaptation is modest or non-existent for most persons until effort approximates 50 to 60% of maximum intensity. Thereafter, gains are rapid until they plateau at the top of the curves, between 85 and 90% of maximal effort, indicating that exercise is too intense or that there is insufficient time for recovery, or both. The dose-response curves shift to the right as physical condition improves. The rate of adaptation varies among individuals, although improvements are generally similar at different ages and for males and females.

Components of the Prescription

The prescription dosage consists of intensity, duration, frequency, and mode of exercise.

Intensity

The single most important factor of the exercise prescription is intensity of effort, usually expressed as a percentage of functional aerobic capacity or maximal heart rate (MHR). There is a strong and consistent correlation between oxygen uptake and heart rate as a percentage of maximum (Fig. A.2), regardless of the level of physical condition, gender, or muscle groups being compared.

Several approaches may be used to prescribe training intensity. In any case maximal exercise testing is recommended for best results. The ACSM Guidelines provide clear recommendations for testing. Heart rate prescriptions based on submaximal testing or age-estimated maximal heart rates have the potential for considerable error and, as a result, may be too strenuous and pose the risk of injury or too easy and, hence, ineffective.

The target heart rate (THR) is ordinarily established between 70 and 90% MHR, approximately 60 to 80% VO_{2max}. Those who are poorly conditioned as well as patients with cardiopulmonary disease can benefit from training at heart rates less than 70% MHR, while competitive athletes may require greater than 90% MHR for training adaptation.

Source: P.S. Fardy and F.G. Yanowitz, *Cardiac Rehabilitation, Adult Fitness, and Exercise Testing,* 3rd ed. © Williams & Wilkins, Baltimore, MD, 1995, pp. 246–247. Used with permission.

1. Identify the headings of the excerpt and the relationship among them. Which headings are primary-level headings; which are secondary; which are tertiary (third-level heads)? Which heading serves as an umbrella for the rest?

   _____

   _____

2. What do the headings tell you about the content of the excerpt?

   _____

   _____

3. Identify the terms with abbreviations after them. What does this tell you about these words? How is the graph in Figure A.1 useful?

   _____

   _____

   _____

4. After reading the chapter headings, write three study questions. List the questions below:

   _____

   _____

   _____

5. Using a marker pen, highlight key phrases and sentences. Write short marginal notes to help you review the material at a later point.

6. After reading this page, list three key concepts that you will need to study:

   a. _____

   b. _____

   c. _____

6.2 *Focusing on Your Purpose for Reading*

Read the box (on the following page) on kinetic and potential energy and the first law of thermodynamics taken from *Life On Earth* by Teresa Audesirk and Gerald Audesirk.[7] When you have finished, answer the questions below.

1. *Reading for critical evaluation.* Evaluate the material by answering these questions:

Were the ideas clearly supported by examples? If you feel one or more were not supported, give an example.

_____
_____
_____
_____

Did the author make any assumptions that weren't examined? If so, name one or more.

_____
_____
_____
_____

Do you disagree with any part of the material? If so, which part, and why?

_____
_____
_____
_____

Do you have any suggestions for how the material could have been presented more effectively?

_____
_____
_____
_____

Among the fundamental characteristics of all living organisms is the ability to guide chemical reactions within their bodies along certain pathways. The chemical reactions serve many functions, depending on the nature of the organism: to synthesize the molecules that make up the organism's body, to reproduce, to move, even to think. Chemical reactions either require or release **energy**, which can be defined simply as *the capacity to do work*, including synthesizing molecules, moving things around, and generating heat and light. In this chapter we discuss the physical laws that govern energy flow in the universe, how energy flow in turn governs chemical reactions, and how the chemical reactions within living cells are controlled by the molecules of the cell itself. Chapters 7 and 8 focus on photosynthesis, the chief "port of entry" for energy into the biosphere, and glycolysis and cellular respiration, the most important sequences of chemical reactions that release energy.

Energy and the Ability to Do Work

As you learned in Chapter 2, there are two types of energy: **kinetic energy** and **potential energy**. Both types of energy may exist in many different forms. Kinetic energy, or *energy of movement*, includes light (movement of photons), heat (movement of molecules), electricity (movement of electrically charged particles), and movement of large objects. Potential energy, or *stored energy*, includes chemical energy stored in the bonds that hold atoms together in molecules, electrical energy stored in a battery, and positional energy stored in a diver poised to spring (Fig. 4-1). Under the right conditions, kinetic energy can be transformed into potential energy, and vice versa. For example, the diver converted kinetic energy of movement into potential energy of position when she climbed the ladder up to the platform; when she jumps off, the potential energy will be converted back into kinetic energy.

To understand how energy flow governs interactions among pieces of matter, we need to know two things: (1) the quantity of available energy and (2) the usefulness of the energy. These are the subjects of the laws of thermodynamics, which we will now examine.

The Laws of Thermodynamics Describe the Basic Properties of Energy

All interactions among pieces of matter are governed by the two **laws of thermodynamics**, physical principles that define the basic properties and behavior of energy. The laws of thermodynamics deal with "isolated systems," which are any parts of the universe that cannot exchange either matter or energy with any other parts. Probably no part of the universe is completely isolated from all possible exchange with every other part, but the concept of an isolated system is useful in thinking about energy flow.

The First Law of Thermodynamics States That Energy Can Neither Be Created nor Destroyed

The **first law of thermodynamics** states that within any isolated system, energy can neither be created nor destroyed, although it can be changed in form (for example, from chemical energy to heat energy). In other words, within an isolated system *the total quantity of energy remains constant*. The first law is therefore often called the law of conservation of energy. To use a familiar example, let's see how the first law applies to driving your car (Fig. 4-2). We can consider that your car (with a full tank of gas), the road, and the surrounding air roughly constitute an isolated system. When you drive your car, you convert the potential chemical energy of gasoline into kinetic energy of movement and heat energy. The total amount of energy that was in the gasoline before it was burned is the same as the total amount of this kinetic energy and heat.

An important rule of energy conversions is this: Energy always flows "downhill," from places with a high concentration of energy to places with a low concentration of energy. This is the principle behind engines. As we described in Chapter 2, temperature is a measure of how fast molecules move. The burning gasoline in your car's engine consists of molecules moving at extremely high speeds: a high concentration of energy. The cooler air outside the engine consists of molecules moving at much lower speeds: a low concentration of energy. The molecules in the engine hit the piston harder than the air molecules outside the engine do, so the piston moves upward, driving the gears that move the car. Work is done. When the engine is turned off, it cools down as heat is transferred from the warm engine to its cooler surroundings. The molecules on both sides of the piston move at the same speed, so the piston stays still. No work is done.

Source: Life on Earth by Audesirk/Audesirk, ©1997. Reprinted by permission of Prentice-Hall, Inc., Upper Saddle River, NJ.

2. *Reading for practical application.* Imagine you have to give a presentation on this material the next time the class meets. On a separate sheet of paper, create an outline or think link that maps out the key elements you would discuss.

3. *Reading for comprehension.* Answer the following questions to determine the level of your comprehension.

Name the two types of energy.

_____

_____

Which one "stores" energy?

_____

_____

_____

_____

Can kinetic energy be turned into potential energy?

_____

_____

_____

_____

What is the term that describes the basic properties and behaviors of energy?

_____

Mark the following statements as true (T) or false (F).

_____ Within any isolated system, energy can be neither created nor destroyed.

_____ Energy always flows downhill, from high-concentration levels to low.

_____ All interactions among pieces of matter are governed by two laws of thermodynamics.

_____ Some parts of the universe are isolated from other parts.

KEY TO SELF-EXPRESSION
Discovery Through Journal Writing

To record your thoughts, use a separate journal or the lined pages at the end of the chapter.

Reading Challenges

What is your most difficult challenge when reading assigned materials? A challenge might be a particular kind of reading material, a reading situation, or the achievement of a certain goal when reading. Considering the tools that this chapter presents, make a plan that addresses this challenge. What techniques might be able to help you most? How and when will you try them out? What positive effects do you anticipate they may have on you?

Reading and Studying CHAPTER 6 **165**

Name _____ Date _____ **Journal**

Journal

Name _____ Date _____

7

Note-Taking and Writing

Harnessing the Power of Words and Ideas

Words, joined to form ideas, are tools that have enormous power. Whether you write an essay, a memo to a supervisor, or a love letter over e-mail, words allow you to take your ideas out of the realm of thought and give them a form that other people can read and consider. You can harness their power for your own. Set a goal for yourself: Strive continually to improve your knowledge of how to use words to construct understandable ideas.

This chapter will teach you the note-taking skills you need to record information successfully. It will show you how to express your written ideas completely and how good writing is linked to clear thinking. In class or at work, taking notes and writing well will help you stand out from the crowd.

In this chapter, you will explore answers to the following questions:

◆ How does taking notes help you?

- Which note-taking system should you use?
- How can you write faster when taking notes?
- Why does good writing matter in science?
- What are the elements of effective writing?
- What is the writing process in science?

How Does Taking Notes Help You?

Notes help you learn when you are in class, doing research, or studying. Because it is virtually impossible to take notes on everything you hear or read, the act of note-taking encourages you to decide what is worth remembering. The positive effects of note-taking include:

- Your notes provide material that helps you study information and prepare for tests.
- When you take notes, you become an active, involved listener and learner.
- Note-taking helps increase your observation skills.
- Notes help you think critically and organize ideas.
- The information you learn in class or lab may not appear in any text; you will have no way to study it without writing it down.
- If it is difficult for you to process information while in class, having notes to read and make sense of later can help you learn.
- Note-taking is a skill for life that you will use on the job and in your personal life.

Recording Information in Class

Your notes have two purposes: First, they should reflect what you heard in class, and second, they should be a resource for studying, writing, or comparing with your text material.

Preparing to Take Class Notes

Taking good class notes depends on good preparation, including the following:

- If your instructor assigns reading on a lecture topic, you may choose to complete the reading before class so that the lecture becomes more of a review than an introduction.
- Use separate pieces of 8 1/2-by-11-inch paper for each class. If you use a three-ring binder, punch holes in papers your instructor hands out and insert them immediately following your notes for that day.
- Take a comfortable seat where you can easily see and hear, and be ready to write as soon as the instructor begins speaking.

- Choose a note-taking system that helps you handle the instructor's speaking style. While one instructor may deliver organized lectures at a normal speaking rate, another may jump from topic to topic or talk very quickly.
- Set up a support system with a student in each class. That way, when you are absent, you can get the notes you missed.

What to Do During Class

Because no one has the time to write down everything he hears, the following strategies will help you choose and record what you feel is important, in a format that you can read and understand later.

- Date each page. When you take several pages of notes during a lecture, add an identifying letter or number to the date on each page: 11/27 A, 11/27 B, ... or 11/27—1 of 3, 11/27—2 of 3.
- Add the specific topic of the lecture at the top of the page. For example:

    11/27 A—Thermal Behavior of Gases

- If your instructor jumps from topic to topic during a single class, try starting a new page for each new topic.
- Ask yourself critical-thinking questions as you listen: Do I need this information? Is the information important or is it just a digression? Is the information fact or opinion? If it is opinion, is it worth remembering?
- Record whatever an instructor emphasizes (see Figure 7-1 for details).
- Continue to take notes during class discussions and question-and-answer periods. What your fellow students ask about may help you as well.

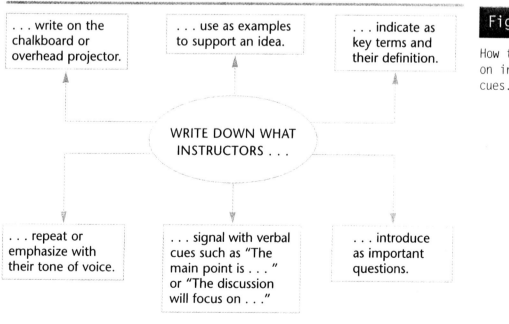

Figure 7-1

How to pick up on instructor cues.

> "Omit needless words. . . . This requires not that the writer make all his sentences short, or that he avoid all detail and treat his subjects only in outline, but that every word tell."
>
> WILLIAM STRUNK, JR.

- Leave one or more blank spaces between points. This white space will help you review your notes, because information will appear in self-contained sections.
- Draw pictures and diagrams that help illustrate ideas.
- Indicate material that is especially important with a star, with underlining, with a highlighter pen, or by writing words in capital letters.
- If you cannot understand what the instructor is saying, leave a space and place a question mark in the margin. Then ask the instructor to explain it again after class or discuss it with a classmate. Fill in the blank when the idea is clear.
- Take notes until the instructor stops speaking. Students who stop writing a few minutes before the class is over can miss critical information.
- Make your notes as legible, organized, and complete as possible. Your notes are only useful if you can read and understand them.

Make Notes a Valuable After-Class Reference

Class notes are a valuable study tool when you review them regularly. To help you do this, try to recopy notes after class. This will help clarify points and assist your memory.

Try to begin your review within a day of the lecture. Read over the notes to learn the information, clarify abbreviations, fill in missing information, and underline or highlight key points. Try to review each week's notes at the end of that week. Think critically about the material, in writing, study group discussions, or quiet reflective thought. You might also try summarizing your notes, either as you review them or from memory.

You can take notes in many ways. Different note-taking systems suit different people and situations. Explore each system and choose what works for you.

WHICH NOTE-TAKING SYSTEM SHOULD YOU USE?

You will benefit most from the system that feels most comfortable to you. As you consider each system, remember the learning styles profile you compiled in Chapter 3. The most common note-taking systems include outlines, the Cornell system, and think links.

Taking Notes in Outline Form

When a reading assignment or lecture seems well organized, you may choose to take notes in outline form. *Outlining* shows the relationships among ideas and their supporting examples through the use of line-by-line phrases set off by varying indentations.

Formal outlines indicate ideas and examples using Roman numerals, capital and lowercase letters, and numbers. When you are pressed for time, such as during class, you can use an informal system of consistent indenting and dashes instead. Formal outlines also require at least two headings on the same level—that is, if you have a II A you must also have a II B. Figure 7-2 shows an outline on impulse and momentum.

Guided Notes

From time to time, an instructor may give you a guide, usually in the form of an outline, to help you take notes in the class. This outline may be on a page that you receive at the beginning of the class, on the board, or on an overhead projector.

Although *guided notes* help you follow the lecture and organize your thoughts during class, they do not replace your own notes. Because they are more of a basic outline of topics than a comprehensive coverage of information, they require that you fill in what they do not cover in detail. If you tune out in class because you think that the guided notes are all you need, you will most likely miss important information.

When you receive guided notes on paper, write directly on the paper if there is room. If not, use a separate sheet and write on it the outline cate-

Figure 7-2

Sample formal outline.

Impulse and Momentum

I. Key Ideas
 A. Impulse and Momentum: Forces that act between colliding objects
  1. Impulse: impulse force = average value of force times time interval
  2. Momentum: vector quantity = body's mass times velocity
 B. Conservation of Momentum: "Law of conservation of momentum"—Acceleration and change in total momentum of system determined by net external force on system.

II. Key Equations
 A. Momentum of a particle $p = mv$
 B. Impulse and Momentum
  1. Impulse $= F_{av} \Delta t = m\Delta v = \Delta(mv) = \Delta p$
  2. Conservation of momentum $= p = p_1 + p_2 =$ constant

Adapted from: Granvil C. Kyker, Jr. (1987). *Study Guide: Paul A Tipler College Physics.* New York: Worth, 81–82.

gories that the guided notes suggest. If the guided notes are on the board or an overhead projector, copy them down, leaving plenty of space in between for your own notes.

Using the Cornell Note-Taking System

The *Cornell note-taking system*, also known as the T-note system, was developed more than 45 years ago by Walter Pauk at Cornell University.[1] The system is successful because it is simple—and because it works. It consists of three sections on ordinary note paper:

- *Section 1*, the largest section, is on the right. Record your notes here in informal outline form.
- *Section 2*, to the left of your notes, is the *cue column*. Leave it blank while you read or listen, then fill it in later as you review. You might fill it with comments that highlight main ideas, clarify meaning, suggest examples, or link ideas and examples. You can even draw diagrams.
- *Section 3*, at the bottom of the page, is the *summary area*, where you summarize the notes on the page. When you review, use this section to reinforce concepts and provide an overview.

When you use the Cornell system, create the note-taking structure before class begins. Picture an upside-down letter T and use Figure 7-3 as your guide. Make the cue column about 2 1/2 inches wide and the summary area 2 inches tall. Figure 7-3 shows how a student used the Cornell system to take notes in an introductory human development course.

Creating a Think Link

A *think link*, also known as a mind map, is a visual form of note-taking. When you draw a think link, you diagram ideas using shapes and lines that link ideas and supporting details and examples. The visual design makes the connections easy to see, and the use of shapes and pictures extends the material beyond just words. Many learners respond well to the power of visualization. You can use think links to brainstorm ideas for paper topics as well.

Visualization
The interpretation of verbal ideas through the use of mental visual images.

One way to create a think link is to start by circling your topic in the middle of a sheet of unlined paper. Next, draw a line from the circled topic and write the name of the first major idea at the end of that line. Circle the idea also. Then jot down specific facts related to the idea, linking them to the idea with lines. Continue the process, connecting thoughts to one another using circles, lines, and words.

A think link may be difficult to construct in class, especially if your instructor talks quickly. In this case, use another note-taking system during class. Then make a think link as you review. Figure 7-4 shows a think link on a sociology concept called *social stratification*.

Once you choose a note-taking system, your success will depend on how well you use it. Personal shorthand will help you make the most of whatever system you choose.

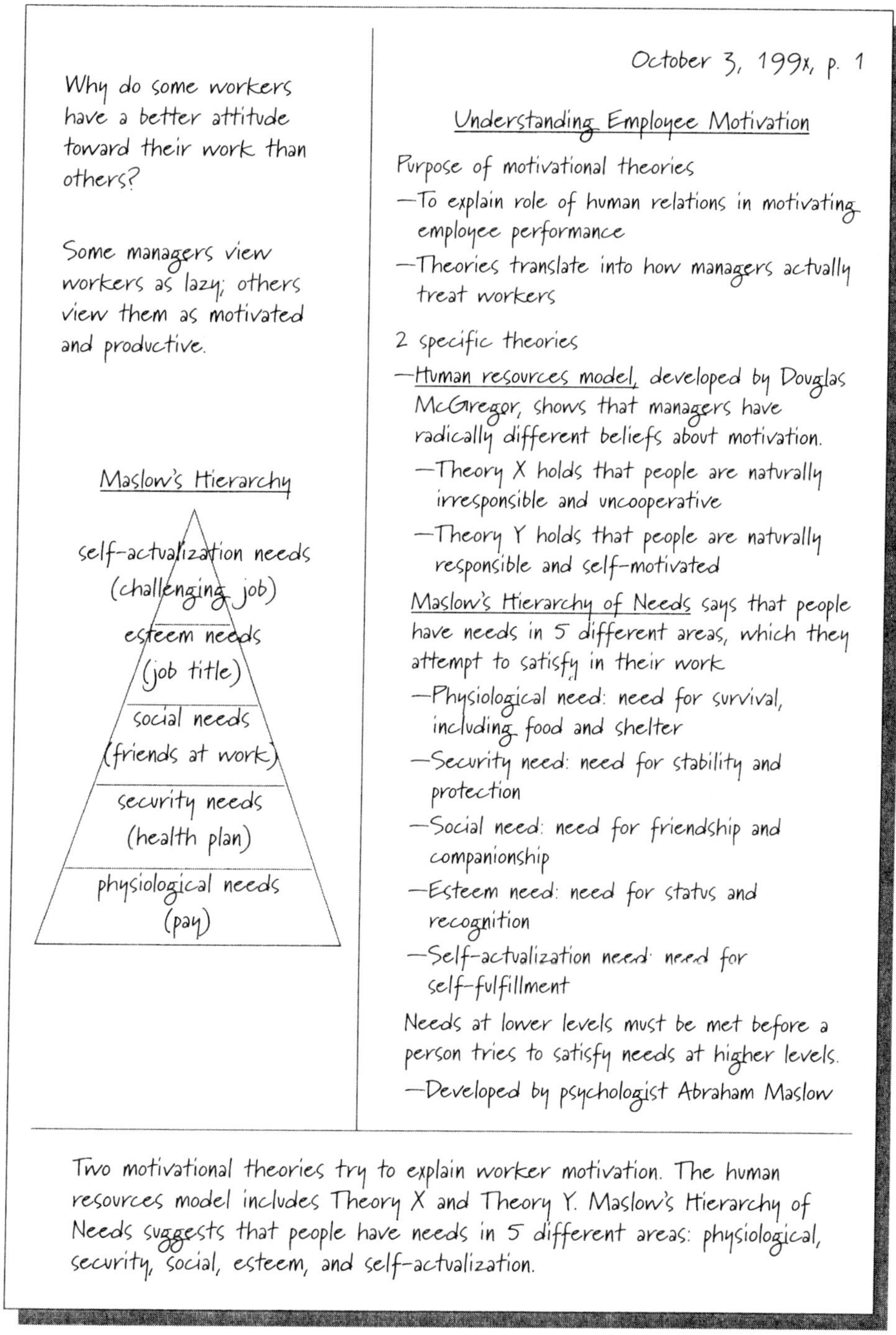

Figure 7-3

Notes taken using the Cornell System.

Why do some workers have a better attitude toward their work than others?

Some managers view workers as lazy; others view them as motivated and productive.

Maslow's Hierarchy

- self-actualization needs (challenging job)
- esteem needs (job title)
- social needs (friends at work)
- security needs (health plan)
- physiological needs (pay)

October 3, 199x, p. 1

Understanding Employee Motivation

Purpose of motivational theories
—To explain role of human relations in motivating employee performance
—Theories translate into how managers actually treat workers

2 specific theories
—Human resources model, developed by Douglas McGregor, shows that managers have radically different beliefs about motivation.
 —Theory X holds that people are naturally irresponsible and uncooperative
 —Theory Y holds that people are naturally responsible and self-motivated

Maslow's Hierarchy of Needs says that people have needs in 5 different areas, which they attempt to satisfy in their work
—Physiological need: need for survival, including food and shelter
—Security need: need for stability and protection
—Social need: need for friendship and companionship
—Esteem need: need for status and recognition
—Self-actualization need: need for self-fulfillment

Needs at lower levels must be met before a person tries to satisfy needs at higher levels.
—Developed by psychologist Abraham Maslow

Two motivational theories try to explain worker motivation. The human resources model includes Theory X and Theory Y. Maslow's Hierarchy of Needs suggests that people have needs in 5 different areas: physiological, security, social, esteem, and self-actualization.

HOW CAN YOU WRITE FASTER WHEN TAKING NOTES?

When taking notes, many students feel they can't keep up with the instructor. Using some personal shorthand (not standard secretarial shorthand) can help to push the pen faster. *Shorthand* is writing that shortens words or replaces

174 CHAPTER 7 Note-Taking and Writing

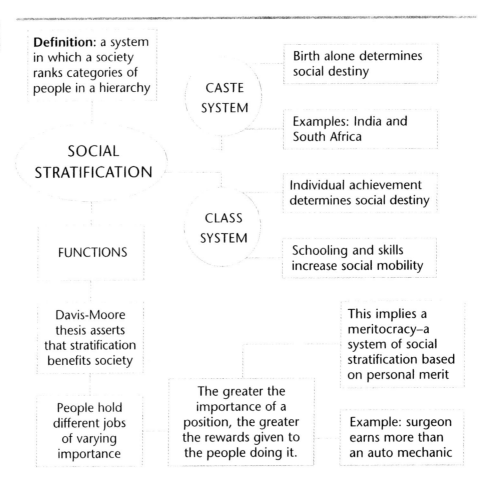

Figure 7-4
Sample think link.

them with symbols. Because you are the only intended reader, you can use symbols and abbreviate words in ways that only you understand.

The only danger with shorthand is that you might forget what your writing means. To avoid this problem, review your shorthand notes while your abbreviations and symbols are fresh in your mind. If there is any confusion, spell out words as you review.

Here are some suggestions that will help you master this important skill:

1. Use the following standard abbreviations in place of complete words:

w/	with	cf	compare, in comparison to
w/o	without	ff	following
→	means; resulting in	Q	question
←	as a result of	p.	page
↑	increasing	*	most importantly
↓	decreasing	<	less than
∴	therefore	>	more than
∵	because	=	equals
≈	approximately	%	percent

+ or &	and	△	change
−	minus; negative	2	to; two; too
No. or #	number	vs	versus; against
i.e.	that is,	e.g.	for example
etc.	and so forth	c/o	care of
ng	no good	lb	pound

2. Shorten words by removing vowels from the middle of words:

   prps = purpose

   Crvtte = Corvette (as on a vanity license plate for a car)

3. Substitute word beginnings for entire words:

   assoc = associate; association

   info = information

4. Form plurals by adding s:

   prblms = problems

   prntrs = printers

5. Make up your own symbols and use them consistently:

   b/4 = before

   2thake = toothache

6. Learn to rely on key phrases instead of complete sentences: "German— nouns capitalized" instead of "In the German language, all nouns are capitalized."

While note-taking focuses on taking in ideas, writing focuses on expressing them. Next you will explore the roles that writing can play in your life.

WHY DOES GOOD WRITING MATTER IN SCIENCE?

Good writing reflects clear thinking. Therefore, a clear thought process is the best preparation for a well-written document, and a well-written document shows the reader a clear thought process. Good writing skills also depend on reading. The more you expose yourself to the work of other writers, the more you will develop your ability to express yourself well. Not only will you learn more words and ideas, but you will also learn about all the different ways a writer can put words together in order to express ideas. In addition, reading generates new ideas inside your mind, ideas you can use in your writing.

In school, almost any course you take will require you to write essays or papers in order to communicate your knowledge and thought process. In order to express yourself successfully in those essays and papers, you need good writing skills. Knowing how to write and express yourself is essential in the workplace as well.

Instructors, supervisors, and other people who see your writing judge your thinking ability based on what you write and how you write it. Over the next few years, you may write papers, essays, answers to essay test questions, job application letters, resumes, business proposals and reports, memos to co-workers, and letters to customers and suppliers. Good writing skills will help you achieve the goals you set out to accomplish with each writing task.

Your writing represents you. In science, good writing presents unique challenges because of the technical nature of scientific language.

WHAT ARE THE ELEMENTS OF EFFECTIVE WRITING?

Every writing situation is different, depending on three elements. Your goal is to understand each element before you begin to write:

- *Your purpose:* What do you want to accomplish with this particular piece of writing?
- *Your topic:* What is the subject about which you will write?
- *Your audience:* Who will read your writing?

Figure 7-5 shows how these elements are interdependent. As a triangle needs three points to be complete, a piece of writing needs these three elements.

Writing Purpose

Writing without having set your purpose first is like driving without deciding where you want to arrive. You'll get somewhere, but chances are it won't be where you needed to go. Therefore, when you write, always define what you want to accomplish before you start.

There are many different purposes for writing. However, the two purposes you will most commonly use in classwork and on the job are to inform and to persuade.

The purpose of *informative writing* is to present and explain ideas. A research paper on how hospitals use donated blood to save lives informs readers without trying to mold opinion. The writer presents facts in an unbiased way, without introducing a particular point of view. Most newspaper

Figure 7-5

The three elements of writing.

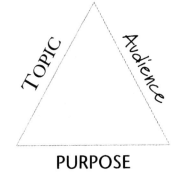

articles, except on the opinion and editorial pages, are examples of informative writing.

Persuasive writing has the purpose of convincing readers that your point of view is correct. Often, persuasive writing seeks to change the mind of the reader. For example, as a member of the student health committee, you write a newspaper column attempting to persuade readers to give blood. Examples of persuasive writing include newspaper editorials, business proposals, and books and magazine articles with a point of view.

Additional possible writing purposes include *entertaining* the reader and *narrating* (describing an image or event to the reader). Although most of your writing in school will inform or persuade, you may occasionally need to entertain or narrate as well. Sometimes purposes will even overlap—you might write an informative essay that entertains at the same time.

Knowing Your Audience

In almost every case, a writer creates written material so that it can be read by others. The two partners in this process are the writer and the audience. Knowing who your audience is will help you communicate successfully.

Key Questions About Your Audience

In school, your primary audience will be your instructors. Instructors use your papers to evaluate the depth of your knowledge. For many assignments, instructors will want you to assume that they are typical readers rather than informed instructors. Writing for typical readers usually means that you should be as complete as possible in your explanations.

At times you may write papers that are intended to address informed instructors or a specific reading audience other than your instructors. In such cases, you may ask yourself some or all of the following questions, depending on which are relevant to your topic.

- What are my readers' ages, cultural backgrounds, interests, and experiences?
- What are their roles? Are they instructors, students, employers, customers?
- How much do they know about my topic? Are they experts in the field or beginners?
- Are they interested, or do I have to convince them to read what I write?
- Can I expect my audience to have an open or closed mind?

After you answer the questions about your audience, take what you have discovered into consideration as you write.

Your Commitment to Your Audience

Your goal is to organize your ideas so that readers can follow them. Suppose, for example, you are writing an informative research paper for a non-expert audience on using on-line services to get a job. One way to accomplish your

Audience
The reader or readers of any piece of written material.

goal is to first explain what these services are and the kinds of help they offer, then describe each service in detail, and finally conclude with how these services will change job hunting in the twenty-first century.

Effective and successful writing involves following the steps of the *writing process*.

What is the Writing Process in Science?

The writing process provides an opportunity for you to state and refine your thoughts until you have expressed yourself as clearly as possible. Critical thinking plays an important role every step of the way. The four main parts of the process are planning, drafting, revising, and editing. Included in this section are the writing steps unique to research writing.

Planning

Planning gives you a chance to think about what to write and how to write it. Planning involves brainstorming for ideas, defining and narrowing your topic by using prewriting strategies, conducting research if necessary, writing a thesis statement, and writing a working outline. Although the steps in preparing to write are listed in sequence, in real life the steps overlap one another as you plan your document.

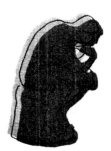

Open Your Mind Through Brainstorming

Whether your instructor assigns a partially defined topic (geneticist Barbara McClintock) or a general category within which you make your own choice (women in science), you should brainstorm to develop ideas about what you want to write. Brainstorming is a creative technique that involves generating ideas about a subject without making judgments. You may want to look at the section on creativity in Chapter 5 for more details.

First, let your mind wander! Write down anything on the assigned subject that comes to mind, in no particular order. Then, organize that list into an outline or think link that helps you see the possibilities more clearly. To make the outline or think link, separate list items into general ideas or categories and sub-ideas or examples. Then associate the sub-ideas or examples with the ideas they support or fit. Figure 7-6 shows a portion of an outline a student, Michael B. Jackson, constructed from his brainstorming list. The assignment is a five-paragraph essay on a life-changing event. Here, only the subject that Michael eventually chose is shown broken down into different ideas.

Narrow Your Topic Through Prewriting Strategies

When your brainstorming has generated some possibilities, you can narrow your topic. Focus on the sub-ideas and examples from your initial brainstorming session. Because they are relatively specific, they will be more likely to point you toward possible topics.

Figure 7-6

Part of a brainstorming outline.

```
A life-changing event...
    —Family
    —Childhood
    →Military
        —travel
        →Boot
            —physical conditioning
                • swim tests
                • intensive training
                • ENDLESS pushups
            —Chief who was our commander
            —mental discipline
                • military lifestyle
                • perfecting our appearance
            —self-confidence
                • walk like you're in control
                • don't blindly accept anything
```

Choose one or more sub-ideas or examples that you like and explore them using prewriting strategies such as brainstorming, freewriting, and asking journalists' questions.[2] Prewriting strategies will help you decide which of your possible topics you would most like to pursue.

Brainstorming. The same process you used to generate ideas will also help you narrow your topic further. Generate thoughts about the possibility you have chosen and write them down. Then, organize them into categories, noticing any patterns that appear. See if any of the sub-ideas or examples seem as if they might make good topics.

Freewriting. Another stream-of-consciousness technique that encourages you to put down ideas on paper as they occur to you is called *freewriting*. When you freewrite, you write whatever comes to mind without censoring your ideas or worrying about grammar, spelling, punctuation, or organization. Freewriting helps you think creatively and gives you an opportunity to begin weaving in information that you know. Freewrite on the sub-ideas or examples you have created to see if you want to pursue any of them. Here is a sample of freewriting:

TERMS
Prewriting strategies Techniques for generating ideas about a topic and finding out how much you already know before you start your research and writing.

> Boot camp for the Coast Guard really changed my life. First of all, I really got in shape. We had to get up every morning at 5 a.m., eat breakfast, and go right into training. We had to do endless military-style push-ups—but we later found out that these have a purpose, to prepare us to hit the deck in the event of enemy fire. We had a lot of aquatic tests, once we were awakened at 3 a.m. to do one in full uniform! Boot camp also helped me to feel confident about myself and be disciplined. Chief Marzloff was the main person who made that happen. He was tough but there was always a reason. He got angry when I used to nod my head whenever he would speak to me, he said that made it seem like I was blindly accepting whatever he said, which was a weakness. From him I have learned to keep an eye on my body's movements when I communicate. I learned a lot more from him too.

Asking journalists' questions. When journalists start working on a story, they ask themselves: Who? What? Where? When? Why? and How? You can use these *journalists' questions* to focus your thinking. Ask these questions about any sub-idea or example to discover what you may want to discuss. For example:

Who? Who was at boot camp? Who influenced me the most?

What? What about boot camp changed my life? What did we do?

When? When in my life did I go to boot camp, and for how long? When did we fulfill our duties?

Where? Where was boot camp located? Where did we spend our day-to-day time?

Why? Why did I decide to go there? Why was it such an important experience?

How? How did we train in the camp? How were we treated? How did we achieve success?

As you prewrite, don't forget to focus on the paper length, due date of your assignment, and any other requirements (such as topic area or purpose). These requirements influence your choice of a final topic. For example, if you had a month to write an informative 20-page paper on learning disabilities, you might choose to discuss the symptoms, diagnosis, effects, and treatment of attention deficit hyperactivity disorder (ADHD). If you were given a week to write a five-page persuasive essay, you might write about how elementary students with ADHD need special training.

Prewriting will help you develop a topic broad enough to give you something with which to work but narrow enough to be manageable. Prewriting also helps you see what you know and what you don't know. If your assignment requires more than you already know, you may need to do research.

Conduct Research

Much of the writing you do in college, such as when you must write a short essay for freshman composition or for an exam, will rely on what you already know about a subject. In these cases, prewriting strategies may generate all the ideas and information you need. In other writing situations, outside sources are necessary. Try doing your research in stages. In the first stage, look for a basic overview that can help you write a thesis statement. In the second stage, go into more depth in your research, tracking down information that will help you fill in the gaps and complete your thoughts.

Write a Thesis Statement

Your work up until this point has prepared you to write a *thesis statement*, the central message you want to communicate. The thesis statement states your subject and point of view, reflects your writing purpose and audience, and acts as the organizing principle of your paper. It tells your readers what they should expect to read. Here is an example from Michael's paper:

Topic	Coast Guard boot camp
Purpose	To inform and narrate
Audience	Instructor with unknown knowledge about the topic
Thesis statement	Chief Marzloff, our Basic Training Company Commander at the U. S. Coast Guard Basic Training Facility, shaped my life through physical conditioning, developing my self-confidence, and instilling strong mental discipline.

A thesis statement is just as important in a short document, such as a letter, as it is in a long paper. For example, when you write a job application letter, a clear thesis statement will help you tell the recruiter why you deserve the job.

Write a Working Outline

The final step in the preparation process involves writing a working outline. Use this outline as a loose guide instead of a finalized structure. As you draft your paper, your ideas and structure may change many times. Only through allowing changes and refinements to happen can you get closer and closer to what you really want to say. Some students prefer a more formal outline structure, while others like to use a think link. Choose whatever form suits you best.

Create a Checklist

Use the checklist in Table 7-1 to make sure your preparation is complete. Under "Date Due," create your own writing schedule, giving each task an intended completion date. Work backwards from the date the assignment is due and estimate how long it will take to complete each step. Refer to Chapter 4 for time management skills that will help you schedule your writing process.

As you develop your schedule, keep in mind that you'll probably move back and forth between tasks. You might find yourself doing two and even three things on the same day. Stick to the schedule as best you can, while bal-

Table 7-1

Preparation checklist.

DATE DUE	TASK	IS IT COMPLETE?
	Brainstorm	
	Define and narrow	
	Use prewriting strategies	
	Conduct research if necessary	
	Write thesis statement	
	Write working outline	
	Complete research	

ancing the other demands of your busy life, and check off your accomplishments on the list as you complete them.

Drafting

"Clear a space for the writing voice . . . you cannot will this to happen. It is a matter of persistence and faith and hard work. So you might as well just go ahead and get started."

ANNE LAMOTT

Some people aim for perfection when they write a first draft. They want to get everything right—from word choice to tone to sentence structure to paragraph organization to spelling, punctuation, and grammar. Resist this tendency because it may lead you to shut the door on ideas before you even know they are there.

A *first draft* involves putting ideas down on paper for the first time—but not the last! You may write many different versions of the assignment until you like what you see. Each version moves you closer to communicating exactly what you want to say in the way you want to say it. The process is like starting with a muddy pond and gradually clearing the mud away until your last version is a clear body of water, showing the rocks and the fish underneath the surface. Think of your first draft as a way of establishing the pond before you start clearing it up.

The elements of writing a first draft are freewriting, crafting an introduction, organizing the ideas in the body of the paper, formulating a conclusion, and citing sources.

Freewriting Your Draft

If the introduction, body, and conclusion are the three parts of the sandwich, freewriting is the process of searching the refrigerator for the ingredients and laying them all out on the table. Take everything that you have developed in the planning stages and freewrite a very rough draft. Don't censor yourself. For now, don't consciously think about your introduction, conclusion, or structure within the paper body. Focus on getting your ideas out of the realm of thought and onto the paper, in whatever form they prefer to be at the moment.

When you have the beginnings of a paper in your hands, you can start to shape it into something with a more definite form. First, work on how you want to begin your paper.

Writing an Introduction

The introduction tells your readers what the rest of the paper will contain. Including the thesis statement is essential. Here, for example, is a draft of an introduction for Michael's paper about the Coast Guard. The thesis statement is underlined at the end of the paragraph:

> Chief Marzloff took on the task of shaping the lives and careers of the youngest, newest members of the U. S. Coast Guard. During my eight weeks in training, he was my father, my instructor, my leader, and my worst enemy. He took his job very seriously and demanded that we do the same. <u>The Chief was instrumental in conditioning our bodies, developing our self-confidence, and instilling mental discipline within us.</u>

When you write an introduction, you might try to draw the reader in with an anecdote—a story that is directly related to the thesis. You can try other hooks, including a relevant quotation, dramatic statistics, and questions that encourage critical thinking. Whatever strategy you choose, be sure it is linked to your thesis statement. In addition, try to state your purpose without referring to its identity as a purpose. For example, in your introductory paragraph, state "Computer technology is infiltrating every aspect of health care," instead of, "In this paper, my purpose is to prove that computer technology is infiltrating every aspect of health care."

After you have an introduction that seems to set up the purpose of your paper, work on making sure the body fulfills that purpose.

Hooks Elements—including facts, quotes, statistics, questions, stories, or statements—that catch the reader's attention and encourage her to want to continue to read.

Creating the Body of a Paper

The body of the paper contains your central ideas and supporting evidence. *Evidence*—proof that informs or persuades—consists of the facts, statistics, examples, and expert opinions that you know or have gathered during research.

Look at the array of ideas and evidences within your draft in its current state. Think about how you might group certain items of evidence with the particular ideas they support. Then, when you see the groups that form, try to find a structure that helps you to organize them into a clear pattern. Here are some strategies to consider.

Arrange ideas by time. Describe events in order or in reverse order.

Arrange ideas according to importance. You can choose to start with the idea that carries the most weight and move to ideas with less value or influence. You can also move from the least important to the most important idea.

Arrange ideas by problem and solution. Start with a specific problem; then discuss one or more solutions.

Writing the Conclusion

Your conclusion is a statement or paragraph that provides closure for your paper. Aim to summarize the information in the body of your paper, as well as to critically evaluate what is important about that information. Try one of the following devices:

- A summary of main points (if material is longer than three pages)
- A story, a statistic, a quote, a question that makes the reader think
- A call to action
- A look to the future

As you work on your conclusion, try not to introduce new facts or restate what you feel you have proved ("I have successfully proven that violent cartoons are related to increased violence in children.") Let your ideas as they are presented in the body of the paper speak for themselves. Readers should feel that they have reached a natural point of completion.

Crediting Authors and Sources

When you write a paper using any materials other than your own thoughts and recollections, the ideas you gathered in your research become part of your own writing. This does not mean that you can claim these ideas as your own or fail to attribute them to someone. You need to credit authors for their ideas and words in order to avoid plagiarism.

To prevent plagiarism, learn the difference between a quotation and a paraphrase. A *quotation* refers to a source's exact words, which are set off from the rest of the text by quotation marks. A *paraphrase* is a restatement of the quotation in your own words, using your own sentence structure. Restatement means to completely rewrite the idea, not just to remove or replace a few words. A paraphrase may not be acceptable if it is too close to the original.

Even an acceptable paraphrase requires a citation of the source of the ideas within it. Take care to credit any source that you quote, paraphrase, or use as evidence. To credit sources, write a footnote or endnote that describes the source. Use the format preferred by your instructor. Writing manuals such as the *APA Publication Manual* contain acceptable formats.

> **TERMS**
>
> **Plagiarism**
> The act of using someone else's exact words, figures, unique approach, or specific reasoning without giving appropriate credit.

Revising

When you *revise*, you critically evaluate the word choice, paragraph structure, and style of your first draft to see how it works. Any draft, no matter how good, can always be improved. Be thorough as you add, delete, replace, and reorganize words, sentences, and paragraphs. You may want to print out your draft and then spend time making notes and corrections on that hard copy before you make changes on a typewritten or computer-printed version. Figure 7-7 on the next page shows a paragraph from Michael's first draft, with revision comments added.

In addition to revising on your own, some of your classes may include peer review (having students read each other's work and offer suggestions). A peer reviewer can tell you what comes across well and what may be confusing. Having a different perspective on your writing is extremely valuable. Even if you don't have an organized peer-review system, you may want to ask a classmate to review your work as a favor to you.

The elements of revision include being a critical writer by checking for clarity and conciseness.

Figure 7-7

Paragraph from first draft.

> Of the changes that ~~happened to us~~ military recruits undergo, the physical transformation is the ~~biggest~~ most evident. ~~When we arrived at the training facility, it was January, cold and cloudy. At the time,~~ Too much ↗ Maybe— upon my January arrival at the training facility, I was a little thin, but I had been working out and thought that I could physically do anything. Oh boy, was I wrong! The Chief said to us right away: "Get down, maggots!" ← his trademark phrase Upon this command, we were all to drop to the ground and do endless military-style push-ups. Water survival tactics were also part of the training ~~that we had to complete~~. Occasionally, my unnecessary dreams of home were interrupted at 3 a.m. when we had a surprise aquatic test. Although we resented ~~didn't feel too happy about~~ this sub-human treatment at the time, we learned to appreciate how the conditioning was turning our bodies into fine-tuned machines.
>
> mention how chief was involved
>
> say more about this (swimming in uniform incident?)

Being a Critical Writer

Critical thinking is as important in writing as it is in reading. Thinking critically when writing will help you move your papers beyond restating what you have researched and learned. Of course, your knowledge is an important part of your writing. What will make your writing even more important and unique, however, is how you use critical thinking to construct your own new ideas and knowledge from what you have learned.

The key to critical writing is asking the question, "So what?" For example, if you were writing a paper on nutrition, you might discuss a variety of good eating habits. Asking "So what?" could lead you into a discussion of *why* these habits are helpful, or what positive effects they have. If you were writing a paper on Aristotle's view of cosmology, you might list the main ideas you noticed he had on the subject. Then, asking "So what?" could lead you to evaluate why Aristotle's ideas are important to our understanding of cosmology today.

As you revise, ask yourself questions that can help you think through ideas and examples, come up with your own original insights about the material, and be as complete and clear as possible. Use the mind actions to guide you. Here are some examples of questions you may ask:

Are these examples clearly connected to an idea or principle?

Are there any similar concepts or facts I know of that can add to how I support this?

What else can I recall that can help to support this idea?

In evaluating any event or situation, have I clearly indicated the causes and effects?

What new principle or idea comes to mind when I think about these examples or facts?

How do I evaluate any effect/fact/situation? Is it good or bad, useful or not?

What different arguments might a reader think of that I should address here?

Finally, critical thinking can help you evaluate the content and form of your paper. As you start your revision, ask yourself the following questions.

- Will my audience understand my thesis and how I've supported it?
- Does the introduction prepare the reader and capture attention?
- Is the body of the paper organized effectively?
- Is each idea fully developed, explained, and supported by examples?
- Are my ideas connected to one another through logical transitions?
- Do I have a clear, concise, simple writing style?
- Does the paper fulfill the requirements of the assignment?
- Does the conclusion provide a natural ending to the paper?

Check for Clarity and Conciseness

Aim to say what you want to say in the clearest, most efficient way possible. A few well-chosen words will do your ideas more justice than a flurry of language. Try to eliminate extra words and phrases. Rewrite wordy phrases in a more concise, conversational way. For example, you can write "if" instead of "in the event that" or "now" instead of "this point in time." "Capriciously, I sauntered forth to the entryway and pummeled the door that loomed so majestically before me" might become "I skipped to the door and knocked loudly."

Editing

In contrast to the critical thinking of revising, *editing* involves correcting technical mistakes in spelling, grammar, punctuation, as well as checking style consistency for elements such as abbreviations and capitalizations. Editing comes last, after you are satisfied with your ideas, organization, and style of writing. If you use a computer, you might want to use the grammar-check and spellcheck functions to find mistakes. A spell-checker helps, but you still need to check your work on your own. While a spell-checker won't pick up the mistake in the following sentence, someone who is reading for sense will:

They are not hear on Tuesdays.

Look also for *sexist language*, which characterizes people based on their gender. Sexist language often involves the male pronoun *he* or *his*. For example "An executive often spends hours each day going through his electronic mail" implies that executives are always men. A simple change will eliminate the sexist language: "Executives often spend hours each day going through their electronic mail," or "An executive often spends hours each day going through his or her electronic mail." Another option is to alternate use of pronouns, using "him" in one section or example and "her" in another. Try to be sensitive to words that leave out or slight women. *Mail carrier* is preferable to *mailman*; *student* to *coed*.

Proofreading is the last stage of editing, occurring when you have a final version of your paper. Proofreading means reading every word and sentence in the final version to make sure they are accurate. Look for technical mistakes, run-on sentences, and sentence fragments. Look for incorrect word usage and references that aren't clear.

Teamwork can be a big help as you edit and proofread, because another pair of eyes may see errors that you didn't notice on your own. If possible, have someone look over your work. Ask for feedback on what is clear and what is confusing. Then ask the reader to edit and proofread for errors.

A Final Checklist

You are now ready to complete your revising and editing checklist. All the tasks listed in Table 7-2 should be complete when you submit your final paper.

Your final paper reflects all the hard work you put in during the writing process. Figure 7-8 shows the final version of Michael's paper.

The Research Format

Written research reports fit into a prescribed formula, making them easily identifiable in the journals you read. The sections are as follows:

Abstract. The beginning of an article is called the abstract. The abstract is approximately a 200-word summary of the entire research study.

Introduction. This section introduces the problem and often gives the research purpose and question, or hypothesis, to be tested in the study.

"See revision as 'envisioning again.' If there are areas in your work where there is a blur or vagueness, you can simply see the picture again and add the details that will bring your work closer to your mind's picture."
NATALIE GOLDBERG

Table 7-2

Revising and editing checklist.

DATE DUE	TASK	IS IT COMPLETE?
	Check the body of the paper for clear thinking and adequate support of ideas	
	Finalize introduction and conclusion	
	Check word spelling and usage	
	Check grammar	
	Check paragraph structure	
	Make sure language is familiar and concise	
	Check punctuation	
	Check capitalization	
	Check transitions	
	Eliminate sexist language	

Figure 7-8

The final version of the paper.

March 19, 1999
Michael B. Jackson

BOYS TO MEN

His stature was one of confidence, often misinterpreted by others as cockiness. His small frame was lean and agile, yet stiff and upright, as though every move were a calculated formula. For the longest eight weeks of my life, he was my father, my instructor, my leader, and my worst enemy. His name is Chief Marzloff, and he had the task of shaping the lives and careers of the youngest, newest members of the U. S. Coast Guard. As our Basic Training Company Commander, he took his job very seriously and demanded that we do the same. Within a limited time span, he conditioned our bodies, developed our self-confidence, and instilled within us a strong mental discipline.

Of the changes that recruits in military basic training undergo, the physical transformation is the most immediately evident. Upon my January arrival at the training facility, I was a little thin, but I had been working out and thought that I could physically do anything. Oh boy, was I wrong! The Chief wasted no time in introducing me to one of his trademark phrases: "Get down, maggots!" Upon this command, we were all to drop to the ground and produce endless counts of military-style push-ups. Later, we found out that exercise prepared us for hitting the deck in the event of enemy fire. Water survival tactics were also part of the training. Occasionally, my dreams of home were interrupted at about 3 a.m. when our company was selected for a surprise aquatic test. I recall one such test that required us to swim laps around the perimeter of a pool while in full uniform. I felt like a salmon swimming upstream,

(continued)

fueled only by natural instinct. Although we resented this sub-human treatment at the time, we learned to appreciate how the strict guidance of the Chief was turning our bodies into fine-tuned machines.

Beyond physical ability, Chief Marzloff also played an integral role in the development of our self-confidence. He would often declare in his raspy voice, "Look me in the eyes when you speak to me! Show me that you believe what you're saying!" He taught us that anything less was an expression of disrespect. Furthermore, he appeared to attack a personal habit of my own. It seemed that whenever he would speak to me individually, I would nervously nod my head in response. I was trying to demonstrate that I understood, but to him, I was blindly accepting anything that he said. He would roar, "That is a sign of weakness!" Needless to say, I am now conscious of all bodily motions when communicating with others. The Chief also reinforced self-confidence through his own example. He walked with his square chin up and chest out, like the proud parent of a newborn baby. He always gave the appearance that he had something to do, and that he was in complete control. Collectively, all of the methods that the Chief used were successful in developing our self-confidence.

Perhaps the Chief's greatest contribution was the mental discipline that he instilled in his recruits. He taught us that physical ability and self-confidence were nothing without the mental discipline required to obtain any worthwhile goal. For us, this discipline began with adapting to the military lifestyle. Our day began promptly at 0500 hours, early enough to awaken the oversleeping roosters. By 0515 hours, we had to have showered, shaved, and perfectly donned our uniforms. At that point, we were marched to the galley for chow, where we learned to take only what is necessary, rather than indulging. Before each meal, the Chief would warn, "Get what you want, but you will eat all that you get!" After making good on his threat a few times, we all got the point. Throughout our stay, the Chief repeatedly stressed the significance of self-discipline. He would calmly utter, "Give a little now, get a lot later." I guess that meant different things to all of us. For me, it was a simple phrase that would later become my personal philosophy on life. The Chief went to great lengths to ensure that everyone under his direction possessed the mental discipline required to be successful in boot camp or in any of life's challenges.

Chief Marzloff was a remarkable role model and a positive influence on many lives. I never saw him smile, but it was evident that he genuinely cared a great deal about his job and all the lives that he touched. This man single-handedly conditioned our bodies, developed our self-confidence, and instilled a strong mental discipline that remains in me to this day. I have not seen the Chief since March 28, 1992, graduation day. Over the years, however, I have incorporated many of his ideals into my life. Above all, he taught us the true meaning of the U. S. Coast Guard slogan, "Semper Peratus" (Always Ready).

Figure 7-8

Continued.

Literature review. The literature review justifies the need for the study by showing how other research on similar subjects leaves gaps in the knowledge base. Sometimes the literature review is not under its own heading but included in the introduction.

Theoretical framework. This is a section heading you will not always see, although sometimes the concepts are woven into the introduction. The theoretical framework is the foundation for the study. For instance, if I was studying health risk behaviors in college students, I might use a framework based on the Health Beliefs Model—a model that describes how a person's beliefs about health affect their health practices.

Sample and setting. This is the section that explains who or what is to be studied, the number studied, and where they/it were studied.

Methods. The research method is described in this section. Here you can see how the researchers went about designing the study, collecting data, and possible statistical tests they plan to use.

Results. The findings of the study are posted here, often in the form of tables or figures with statistical techniques applied.

Discussion. The summary of the entire study often includes recommendations for future research on the subject. This can be a good place to look for ideas for studies of your own.

Suà

Suà is a Shoshone Indian word, derived from the Uto-Aztecna language, meaning "think." While much of the American Indian tradition focuses on oral communication, written languages have allowed American Indian perspectives and ideas to be understood by readers outside the American Indian culture. The writings of Leslie Marmon Silko, J. Scott Momaday, and Sherman Alexie have expressed important insights that all readers can consider.

Think of *suà*, and of how thinking can be communicated to others through writing, every time you begin to write. The power of writing allows you to express your own insights so that others can read them and perhaps benefit from knowing them. Explore your thoughts, sharpen your ideas, and remember the incredible power of the written word.

Chapter 7   Applications

Name _____   Date _____

KEY INTO YOUR LIFE
Opportunities to Apply What You Learn

 Evaluate Your Notes

Choose one particular class period from the last two weeks. Have a classmate photocopy his notes from that class period for you. Then evaluate your notes by comparing them with your classmate's. Think about:

- Do your notes make sense?
- How is your handwriting?
- Do the notes cover everything that was brought up in class?
- Are there examples to back up ideas?
- What note-taking system is used?
- Will these notes help you study?

Write your evaluation here: _____
_____
_____
_____
_____

What ideas or techniques from your classmate's notes do you plan to use in the future? _____
_____
_____
_____
_____

7.2 Class vs. Reading

Pick a class for which you have a regular textbook. Choose a set of class notes on a subject that is also covered in that textbook. Read the textbook section that corresponds to the subject of your class notes, taking notes as you go. Compare your reading notes to the notes you took in class.

Did you use a different system with the textbook or the same system as in class? Why? Which notes can you understand better? Why do you think that's true? _____

_____

_____

_____

_____

What did you learn from your reading notes that you want to bring to your class note-taking strategy? _____

_____

_____

_____

_____

7.3 Prewriting

Choose a topic you are interested in and know something about—for example, college sports, handling stress in a stressful world, our culture's emphasis on beauty and youth, or child rearing. Narrow your topic; then use the following prewriting strategies to discover what you already know about the topic and what you would need to learn if you had to write an essay about the subject for one of your classes. (If necessary, continue this prewriting exercise on a separate sheet of paper.)

Brainstorm your ideas: _____

_____

Freewrite: _____

_____

_____

_____

_____

_____

Ask journalists' questions: _____

_____

_____

_____

7.4 Writing a Thesis Statement

Write two thesis statements for each of the following topics. The first statement should try to inform the reader, while the second should try to persuade. In each case, writing a thesis statement will require that you narrow the topic.

- *The rising cost of health care*

    Thesis with an informative purpose: _____

    _____

    _____

    Thesis with a persuasive purpose: _____

    _____

    _____

- *Taking care of your body and mind*

    Thesis with an informative purpose: _____

    _____

    _____

Thesis with a persuasive purpose: _____

◆ *Career choice*

Thesis with an informative purpose: _____

Thesis with a persuasive purpose: _____

KEY TO SELF-EXPRESSION
Discovery Through Journal Writing

To record your thoughts, use a separate journal or the lined pages at the end of the chapter.

Your Relationship with Words

Some people love to work with words—writing them, reading them, speaking them—while others would rather do anything else. Do you enjoy writing in school, or does writing intimidate you? Do you write anything outside of school? Discuss how you feel about writing.

Name _____ Date _____   **Journal**

Journal

Name _____ Date _____

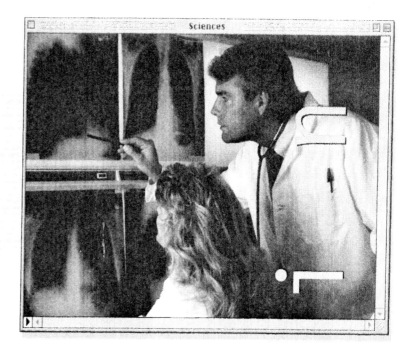

Listening, Memory, and Test Taking

Taking In, Retaining, and Demonstrating Knowledge

College exposes you daily to facts, opinions, and ideas. It is up to you to take in, retain, and demonstrate knowledge of what you learn, for use in or out of school. You can accomplish these goals through active listening, focused use of your memory skills, and thorough preparation for taking tests.

Listening is one of the primary ways of taking in information. Memory skills can help you retain what you've listened to so that you can recall it for a paper, a discussion, or a test. After you've listened and remembered, test taking is your key to demonstrating what you have learned to your instructor or others. In this chapter, you will learn strategies to improve your ability to take in, remember, and show knowledge of what you have learned.

In this chapter, you will explore answers to the following questions:

- How can you become a better listener?
- How does memory work?
- How can you improve your memory?
- How can tape recorders help you listen, remember, and study?
- How can preparation help improve test scores?
- What strategies can help you succeed on written tests?
- How can you learn from test mistakes?

HOW CAN YOU BECOME A BETTER LISTENER?

The act of hearing isn't quite the same as the act of listening. While *hearing* refers to sensing spoken messages from their source, *listening* involves a complex process of communication. Successful listening results in the speaker's intended message reaching the listener. In school, and at home, poor listening results in communication breakdowns and mistakes, and skilled listening promotes progress and success.

Ralph G. Nichols, a pioneer in listening research, studied 200 students at the University of Minnesota over a nine-month period. His findings, summarized in Table 8-1, demonstrate that effective listening depends as much on attitude as on specific skills.[1]

Listening is a teachable—and learnable—skill. Improving your learning skills involves managing listening challenges and becoming an active listener. Although becoming a better listener will help in every class, it is especially important in subject areas that are difficult for you.

Manage Listening Challenges

Classic studies have shown that immediately after listening, students are likely to recall only half of what was said. This is partly due to particular listening challenges, including divided attention and distractions, the tendency to shut out the message, the inclination to rush to judgment, and partial hearing loss or learning disabilities.[2] To help create a positive listening environment, in both your mind and your surroundings, explore how to manage these challenges.

Divided Attention and Distractions

Internal and external distractions often divide your attention. *Internal distractions* include anything from hunger to headache to personal worries. Something the speaker says may also trigger a recollection that may cause your mind to drift. In contrast, *external distractions* include noises (whispering,

Table 8-1. What helps and hinders listening.

LISTENING IS HELPED BY...	LISTENING IS HINDERED BY...
making a conscious decision to work at listening; viewing difficult material as a listening challenge.	caring little about the listening process; tuning out difficult material.
fighting distractions through intense concentration	refusing to listen at the first distraction.
continuing to listen when a subject is difficult or dry, in the hope that one might learn something interesting.	giving up as soon as one loses interest.
withholding judgment until hearing everything.	becoming preoccupied with a response as soon as a speaker makes a controversial statement.
focusing on the speaker's theme by recognizing organizational patterns, transitional language, and summary statements.	getting sidetracked by unimportant details.
adapting note-taking style to the unique style and organization of the speaker.	always taking notes in outline form, even when a speaker is poorly organized, leading to frustration.
pushing past negative emotional responses and forcing oneself to continue to listen.	letting an initial emotional response shut off continued listening.
using excess thinking time to evaluate, summarize, and question what one just heard and anticipating what will come next.	thinking about other things and, as a result, missing much of the message.

honking horns, screaming sirens) and even excessive heat or cold. It can be hard to listen in an overheated room that is putting you to sleep.

Your goal is to reduce distractions and focus on what you're hearing. Sitting where you can see and hear clearly will help you listen well. Dress comfortably and try not to go to class hungry or thirsty.

Shutting Out the Message

Instead of paying attention to everything the speaker says, many students fall into the trap of focusing on specific points and shutting out the rest of the message. Creating a positive listening environment includes accepting responsibility for listening. While the instructor communicates information to you, she cannot force you to listen. You are responsible for taking in that information. Instructors often cover material from outside the textbook during class and then test on that material. If you work to take in the whole message in class, you will be able to read over your notes later and think critically about what is most important.

The Rush to Judgment

People tend to stop listening when they hear something they don't like. If you rush to judge what you've heard, your focus turns to your personal reaction rather than the content of the speaker's message. Judgments also involve reactions to the speakers themselves. If you do not like your instructors or if you have preconceived notions about their ideas or cultural background, you may decide that their words have little value.

Work to recognize and control your judgments (see Chapter 5). Being aware of what you tend to judge will help you avoid putting up a barrier against incoming messages that clash with your opinions or feelings. Try to see education as a continuing search for evidence, regardless of whether it supports or negates your point of view.

Partial Hearing Loss and Learning Disabilities

"No one cares to speak to an unwilling listener. An arrow never lodges in a stone; often it recoils upon the sender of it."

ST. JEROME

Good listening techniques don't solve every listening problem. Students who have a partial hearing loss have a physical explanation for why listening is difficult. If you have some level of hearing loss, seek out special services that can help you listen in class. You may require special equipment or might benefit from tutoring. You may be able to arrange to meet with your instructor outside of class to clarify your notes.

Other diagnosed disabilities, such as attention deficit hyperactivity disorder (ADHD) or a problem with processing heard language, can cause difficulties with both focusing on and understanding that which is heard. People with such disabilities have varied ability to compensate for and overcome them. If you have or suspect you have, a disability, don't blame yourself for having trouble. Your counseling center, student health center, advisor, and instructors should be able to give you particular assistance in working through your challenges or information on getting tested.

How DOES MEMORY WORK?

You need an effective memory in order to use the knowledge you take in throughout your life. Human memory works like a computer. Both have essentially the same purpose: to encode, store, and retrieve information.

During the *encoding stage*, information is changed into usable form. On a computer, this occurs when keyboard entries are transformed into electronic symbols and stored on a disk. In the brain, sensory information becomes impulses that the central nervous system reads and codes. You are encoding, for example, when you study a list of chemistry formulas.

During the *storage stage*, information is held in memory (the mind's version of a computer hard drive) for later use. In this example, after you complete your studying of the formulas, your mind stores them until you need to use them.

During the *retrieval stage*, memories are recovered from storage by recall, just as a saved computer program is called up by name and used again. In this example, your mind would retrieve the chemistry formulas when you had to take a test or solve a problem.

Memories are stored in three different storage banks. The first, called *sensory memory*, is an exact copy of what you see and hear, and it lasts for a second or less. Certain information is then selected from sensory memory and moves into *short-term memory*, a temporary information storehouse that lasts no more than 10 to 20 seconds. You are consciously aware of material in your short-term memory. While unimportant information is quickly dumped, important information is transferred to *long-term memory*—the mind's more permanent storehouse.

Having information in long-term memory does not mean that you will be able to recall it when needed. Particular techniques can help you improve your recall.

How can you improve your memory?

Your anatomy instructor is giving a test tomorrow on the parts of the brain. You feel confident, because you spent hours last week memorizing the material. Unfortunately, by the time you take the test, you may remember very little. That's because most forgetting occurs within minutes after memorization.

In a classic study conducted in 1885, researcher Herman Ebbinghaus memorized a list of meaningless three-letter words such as CEF and LAZ. Within one short hour, he measured that he had forgotten more than 50 percent of what he learned. After two days, he knew fewer than 30 percent. Although his recall of the syllables remained fairly stable after that, the experiment shows how fragile memory can be, even when you take the time and energy to memorize information.[3]

People who have superior memories may have an inborn talent for remembering. More often, though, they have mastered techniques for improving recall. Remember that techniques aren't a cure-all for memory difficulties, especially for those who may have learning disabilities. If you have a disability, the following memory techniques may help you but may not be enough. Seek specific assistance if you consistently have trouble remembering.

Memory Improvement Strategies

As a student, your job is to understand, learn, and remember information, from general concepts to specific details. The following suggestions will help improve your recall.

Develop a Will to Remember

Why can you remember the lyrics to dozens of popular songs but not the functions of the pancreas? Perhaps this is because you want to remember them, you connect them with a visual image, or you have an emotional tie to them. To achieve the same results at school or on the job, tell yourself that what you are learning is important and that you need to remember it. Saying these words out loud can help you begin the active, positive process of memory improvement.

Understand What You Memorize

Make sure that everything you want to remember makes sense. Something that has meaning is easier to recall than something that is gibberish. This basic principle applies to everything you study—from biology and astronomy to history and English literature.

Recite, Rehearse, and Write

When you *recite* material, you repeat it aloud in order to remember it. Reciting helps you retrieve information as you learn it and is a crucial step in studying (see Chapter 6). *Rehearsing* is similar to reciting, but is done in silence, in your mind. It involves the process of repeating, summarizing, and associating information with other information. *Writing* is rehearsing on paper. The act of writing solidifies the information in your memory.

Separate Main Points from Unimportant Details

If you use critical-thinking skills to select and focus on the most important information, you can avoid overloading your mind with extra clutter. To focus on key points, highlight only the most important information in your texts and write notes in the margins about central ideas. When you review your lecture notes, highlight or rewrite the most important information to remember.

Study During Short but Frequent Sessions

Research shows that you can improve your chances of remembering material if you learn it more than once, which is why "cramming" doesn't work. To get the most out of your study sessions, spread them over time. A pattern of short sessions followed by brief periods of rest is more effective than continual studying with little or no rest. Even though you may feel as though you accomplish a lot by studying for an hour without a break, you'll probably remember more from three 20-minute sessions. Try sandwiching study time into breaks in your schedule, such as when you have time between classes.

Separate Material into Manageable Sections

When material is short and easy to understand, studying it from start to finish may work. For longer material, you may benefit from dividing it into logical sections, mastering each section, putting all the sections together, and then testing your memory of all the material. Actors take this approach when learning the lines of a play, and it can work just as well for students.

Use Visual Aids

Any kind of visual representation of study material can help you remember. You may want to convert material into a think link or outline. Write material in any visual shape that helps you recall it and link it to other information.

Flashcards are a great visual memory tool. They give you short, repeated review sessions that provide immediate feedback. Make them from 3-by-5-inch index cards. Use the front of the card to write a word, idea, or phrase you want to remember. Use the back side for a definition, explanation, and other key facts. Figure 8-1 shows two flashcards for studying psychology.

Here are some additional suggestions for making the most of your flashcards:

- *Use the cards as a self-test.* Divide the cards into two piles: the material you know and the material you are learning. You may want to use rubber bands to separate the piles.
- *Carry the cards with you and review them frequently.* You'll learn the most if you start using cards early in the course, well ahead of exam time.
- *Shuffle the cards and learn information in various orders.* This will help avoid putting too much focus on some information and not enough on others.
- *Test yourself in both directions.* First, look at the terms or ideas and provide definitions or explanations. Then turn the cards over and reverse the process.

Mnemonic Devices

Certain show business performers entertain their audiences by remembering the names of 100 strangers or flawlessly repeating 30 ten-digit phone numbers. These performers probably have superior memories, but genetics alone can't produce these results. They also rely on memory techniques, known as mnemonic devices (pronounced neh-MAHN-ick) to help them.

Mnemonic devices work by connecting information you are trying to learn with simpler information or information that is familiar. Instead of

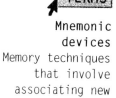

TERMS

Mnemonic devices
Memory techniques that involve associating new information with information you already know.

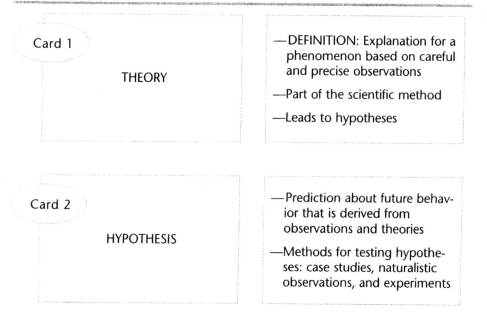

Figure 8-1

Flashcards help you memorize important facts.

learning new facts by rote (repetitive practice), associations give you a hook on which to hang these facts and retrieve them. Mnemonic devices make information familiar and meaningful through unusual, unforgettable mental associations and visual pictures.

Here's an example to prove the power of mnemonics. Suppose you want to remember the names of the first six presidents of the United States. The first letters of their last names—Washington, Adams, Jefferson, Madison, Monroe, and Adams—together read W A J M M A. To remember them, you might add an "e" after the "J" and create a short nonsense word, "wajemma." To remember their first names—George, John, Thomas, James, James, and John—you might set the names to the tune of "Happy Birthday" or any musical tune that you know.

Visual images and acronyms are a few of the more widely used kinds of mnemonic devices. Apply them to your own memory challenges.

Create Visual Images and Associations

Visual images are easier to remember than images that rely on words alone. In fact, communication through visual images goes back to the prehistoric era when people made drawings, which still exist, on cave walls. It's no accident that the phrase "a picture is worth a thousand words" is so familiar. The best mental pictures often involve colors, three-dimensional images, action scenes, and disproportionate, funny, or ridiculous images. Especially for visual learners, turning information into mental pictures helps improve memory.

Create Acronyms

Another helpful association method involves the use of the acronym. The acronym "Roy G. Biv" often helps students remember the colors of the spectrum. Roy G. Biv stands for Red, Orange, Yellow, Green, Blue, Indigo, Violet. In history, you can remember the "big-three" Allies during World War II—Britain, America, and Russia—with the acronym BAR.

When you can't create a name like Roy G. Biv, create an acronym from an entire sentence, in which the first letter of each word in the sentence stands for the first letter of each memorized term. When science students want to remember the list of planets in order of their distance from the sun, they learn the sentence: My very elegant mother just served us nine pickles (Mercury, Venus, Earth, Mars, Jupiter, Saturn, Uranus, Neptune, and Pluto).

Improving your memory requires energy, time, and work. In school, it also helps to master PQ3R, the textbook study technique that was introduced in Chapter 6. By going through the steps in PQ3R and using the specific memory techniques described in this chapter, you will be able to learn more in less time—and remember what you learn long after exams are over.

How can tape recorders help you listen, remember, and study?

The selective use of a tape recorder can provide helpful backup to your listening and memory skills and to your study materials. It's important, though, not to let tape recording substitute for active participation. Not all students

"The true art of memory is the art of attention."

SAMUEL JOHNSON

Acronym
A word formed from the first letters of a series of words, created in order to help you remember the series.

like to use tape recorders, but if you choose to do so, here are some guidelines and a discussion of potential effects.

Guidelines for Using Tape Recorders

Ask the instructor whether he permits tape recorders in class. Some instructors don't mind, while others don't allow students to use them.

Use a small, portable tape recorder. Sit near the front for the best possible recording.

Participate actively in class. Take notes just as you would if the tape recorder were not there.

Use tape recorders to make study tapes. Questions on tape can be like audio flashcards. One way to do it is to record study questions, leaving ten to fifteen seconds between questions for you to answer out loud. Recording the correct answer after the pause will give you immediate feedback. For example, part of a recording for a writing class might say, "The three elements of effective writing are . . . (10–15 seconds) . . . topic, audience, and purpose."

Potential Positive Effects of Using Tape Recorders

- You can listen to an important portion of the lecture over and over again.
- You can supplement or clarify sections of the lecture that confused you or that you missed.
- Tape recordings can provide additional study materials to listen to when you exercise or drive in your car.
- Tape recordings can help study groups reconcile conflicting notes.
- If you miss class, you might be able to have a friend record the lecture for you.

Potential Negative Effects of Using Tape Recorders

- You may tend to listen less in class.
- You may take fewer notes, figuring that you will rely on your tape.
- It may be time-consuming. When you attend a lecture in order to record it and then listen to the entire recording, you have taken twice as much time out of your schedule.
- If your tape recorder malfunctions or the recording is hard to hear, you may end up with very little study material, especially if your notes are sparse.

Think critically about whether using a tape recorder is a good idea for you. If you choose to try it, let the tape recorder be an additional resource for you instead of a replacement for your active participation and skills. Tape-recorded lectures and study tapes are just one study resource you can use in preparation for the tests that will often come your way.

How Can Preparation Help Improve Test Scores?

Many people don't look forward to taking tests. If you are one of those people, try thinking of exams as preparation for life. When you volunteer, get a job, or work on your family budget, you'll have to apply what you know. This is exactly what you do when you take a test.

Like a runner who prepares for a marathon by exercising, eating right, taking practice runs, and getting enough sleep, you can take steps to master your exams. Your first step is to study until you know the material that will be on the test. Your next step is to use the following strategies to become a successful test taker: Identify test type and material covered, use specific study skills, prepare physically, and conquer test anxiety.

Identify Test Type and Material Covered

Before you begin studying, try to determine the type of test and what it will cover:

- Will it be a short-answer test with true/false and multiple-choice questions, an essay test, or a combination?
- Will the test cover everything you studied since the semester began or will it be limited to a narrow topic?
- Will the test be limited to what you learned in class and in the text or will it also cover outside readings?
- Will the test be written or practical for a lab class?

Your instructors may answer these questions for you. Even though they may not tell you the specific questions that will be on the test, they might let you know what blocks of information will be covered and the question format. Some instructors may even drop hints throughout the semester about possible test questions. While some comments are direct ("I might ask a question on the subject of _____ on your next exam"), other clues are subtle. For example, when instructors repeat an idea or when they express personal interest in a topic ("One of my favorite theories is . . ."), they are letting you know that the material may be on the test.

Here are a few other strategies for predicting what may be on a test:

Use PQ3R to identify important ideas and facts. Often, the questions you write and ask yourself when you read assigned materials may be part of the test. In addition, any textbook study questions are good candidates for test material.

If you know people who took the instructor's course before, ask them about class tests. Try to find out how difficult the tests are and whether the test focuses more on assigned readings or class notes. Ask about instructor preferences. If you learn that the instructor pays close attention to detail such as facts or grammar, plan your work accordingly.

- Examine old tests if instructors make them available in class or on reserve in the library. If you can't get copies of old tests, use clues from the class to predict test questions. After taking the first exam in the course, you will have a lot more information about what to expect in the future.

   If the test is a practical lab exam find out if you can practice ahead of time. For instance, if you will be tested on plant identification, ask your instructor how to find samples for test preparation.

Use Specific Study Skills

Certain study skills are especially useful for test taking. They include choosing study materials, setting a study schedule, critical thinking, and taking a pretest.

Choose Study Materials

Once you have identified as much as you can about the subject matter of the test, choose the materials that contain the information you need to study. You can save yourself time by making sure that you aren't studying anything you don't need to. Go through your notes, your texts, any primary source materials that were assigned, and any handouts from your instructor. Set aside any materials you don't need so that they don't take up your valuable time.

Set a Study Schedule

- Use your time management skills to set a schedule that will help you feel as prepared as you can be. Consider all the relevant factors—the materials you need to study, how many days or weeks until the test date, and how much time you can study each day. If you establish your schedule ahead of time and write it in your date book, you will be much more likely to follow it.

   Schedules will vary widely according to situation. For example, if you have only three days before the test and no other obligations during that time, you might set two 2-hour study sessions for yourself during each day. On the other hand, if you have two weeks before a test date, classes during the day, and work three nights a week, you might spread out your study sessions over the nights you have off work during those two weeks.

Prepare Through Critical Thinking

- Using the techniques from Chapter 5, approach your test preparation critically, working to understand rather than just to pass the test by repeating facts. As you study, try to connect ideas to examples, analyze causes and effects, establish truth, and look at issues from different perspectives. Although it takes work, critical thinking will promote a greater understanding of the subject and probably a higher grade on the exam. Using critical thinking is especially important for essay tests. Prepare by identifying potential essay questions and writing your responses.

Take a Pretest

Use questions from the ends of textbook chapters to create your own pretest. Choose questions that are likely to be covered on the test, then answer them under testlike conditions—in quiet, with no books or notes to help you, and with a clock telling you when to quit. Try to duplicate the conditions of the actual test. If your course doesn't have an assigned text, develop questions from your notes and from assigned outside readings.

Prepare Physically

When taking a test, you often need to work efficiently under time pressure. If your body is tired or under stress, you will probably not think as clearly or perform as well. Avoid pulling an all-nighter. Get some sleep so that you can wake up rested and alert. If you are one of the many who press the snooze button in their sleep, try setting two alarm clocks and placing them across the room from your bed. That way you'll be more likely to get to your test on time.

Eating right is also important. Sugar-laden snacks will bring your energy up only to send it crashing back down much too soon. Similarly, too much caffeine can add to your tension and make it difficult to focus. Eating nothing will leave you drained, but too much food can make you want to take a nap. The best advice is to eat a light, well-balanced meal before a test. When time is short, grab a quick-energy snack such as a banana, some orange juice, or a granola bar. Sleep, protein, and carbs will help your recall. Being tired, or wired (on caffeine) will hinder recall.

Conquer Test Anxiety

A certain amount of stress can be a good thing. Your body is on alert, and your energy motivates you to do your best. For many students, however, the time before and during an exam brings a feeling of near-panic known as *test anxiety*. Described as a bad case of nerves that makes it hard to think or remember, test anxiety can make your life miserable and affect how you do on tests. When anxiety blocks performance, here are some suggestions:

Prepare so you'll feel in control. The more you know about what to expect on the exam, the better you'll feel. Find out what material will be covered, the format of the questions, the length of the exam, and the percentage of points assigned to each question.

Put the test in perspective. No matter how important it may seem, a test is only a small part of your educational experience and an even smaller part of your life. Your test grade does not reflect the kind of person you are or your ability to succeed in life.

Make a study plan. Divide the plan into a series of small tasks. As you finish each one, you'll feel a sense of accomplishment and control.

Practice relaxation. When you feel test anxiety coming on, take some deep breaths, close your eyes, and visualize positive mental images related to the test, like getting a good grade and finishing confidently with time to spare.

Test Anxiety and the Returning Adult Student

If you're returning to school after five, ten, or even twenty years, you may wonder if you can compete with younger students or if your mind is still able to learn new material. To counteract these feelings of inadequacy, focus on how your life experiences have given you skills you can use. For example, managing work and a family requires strong time management, planning, and communication skills that can help you plan your study time, juggle school responsibilities, and interact with students and instructors.

In addition, your life experiences give you examples with which you can understand ideas in your courses. For example, your relationship experiences may help you understand concepts in a psychology course; managing your finances may help you understand math; and work experience may give you a context for learning teamwork. If you recognize and focus on your knowledge and skills, you may improve your ability to achieve your goals.

WHAT STRATEGIES CAN HELP YOU SUCCEED ON WRITTEN TESTS?

Even though every test is different, there are general strategies that will help you handle almost all tests, including short-answer and essay exams.

Write Down Key Facts

Before you even look at the test, write down any key information—including formulas, rules, and definitions—that you studied recently or even right before you entered the test room. Use the back of the question sheet or a piece of scrap paper for your notes (make sure it is clear to your instructor that this scrap paper didn't come into the test room already filled in!). Recording this information right at the start will make forgetting less likely.

Begin With an Overview of the Exam

Even though exam time is precious, spend a few minutes at the start of the test to get a sense of the kinds of questions you'll be answering, what kind of thinking they require, the number of questions in each section, and the point value of each section. Use this information to schedule the time you spend on each section. For example, if a two-hour test is divided into two sections of equal point value—an essay section with four questions and a short-answer section with sixty questions—you can spend an hour on the essays (fifteen minutes per question) and an hour on the short-answer section (one minute per question).

As you make your calculations, think about the level of difficulty of each section. If you think you can handle the short-answer questions in less than an hour and that you'll need more time with the essays, rebudget your time in a way that works for you.

Know the Ground Rules

A few basic rules apply to any test. Following them will give you an advantage.

Read test directions. While a test made up of 100 true/false questions and one essay may look straightforward, the directions may tell you to answer eighty, or that the essay is a optional bonus. Some questions or sections may be weighted more heavily than others. Try circling or underlining key words and numbers that remind you of the directions.

Begin with the parts or questions that seem easiest to you. Starting with what you know best can boost your confidence and help you save time to spend on the harder parts.

Watch the clock. Keep track of how much time is left and how you are progressing. You may want to plan your time on a scrap piece of paper, especially if you have one or more essays to write. Wear a watch or bring a small clock with you to the test room. A wall clock may be broken, or there may be no clock at all! Also, take your time. Rushing is almost always a mistake, even if you feel you've done well. Stay till the end so you can refine and check your work.

Master the art of intelligent guessing. When you are unsure of an answer, you can leave it blank or you can guess. In most cases, guessing will benefit you. First eliminate all the answers you know—or believe—are wrong. Try to narrow your choices to two possible answers; then, choose the one that makes more sense to you. When you recheck your work, decide if you would make the same guesses again, making sure there isn't a qualifier or fact that you hadn't noticed before.

Follow directions on machine-scored tests. Machine-scored tests require that you use a special pencil to fill in a small box on a computerized answer sheet. Use the right pencil (usually a number 2) and mark your answer in the correct space. Neatness counts on these tests, because the computer can misread stray pencil marks or partially erased answers. Periodically, check the answer number against the question number to make sure they match. One question skipped can cause every answer following it to be marked incorrectly.

Use Critical Thinking to Avoid Errors

When the pressure of a test makes you nervous, critical thinking can help you work through each question thoroughly and avoid errors. Following are some critical-thinking strategies to use during a test.

TERMS

Qualifier
A word, such as *always, never,* or *often,* that changes the meaning of another word or word group.

Recall facts, procedures, rules, and formulas. You base your answers on the information you recall. Think carefully to make sure you recall it accurately.

Think about similarities. If you don't know how to attack a question or problem, consider any similar questions or problems that you have worked on in class or while studying.

Notice differences. Especially with objective questions, items that seem different from what you have studied may indicate answers you can eliminate.

Think through causes and effects. For a numerical problem, think through how you plan to solve it and see if the answer—the effect of your plan—makes sense. For an essay question that asks you to analyze a condition or situation, consider both what caused it and what effects it has.

Find the best principle to match the example or examples given. For a numerical problem, decide what formula (principle) best applies to the example or examples (the data of the problem). For an essay question, decide what principle applies to, or links, the examples given.

Support principles with examples. When you put forth a principle, or idea, in an answer to an essay question, be sure to back up your idea with an adequate number of examples that fit.

Evaluate each test question. In your initial approach to any question, evaluate what kinds of thinking will best help you solve it. For example, essay questions often require cause and effect and principle to example thinking, while objective questions often benefit from thinking through similarities and differences.

The general strategies you have just explored also can help you address specific types of test questions.

Master Different Types of Test Questions

Although the goal of all test questions is to discover how much you know about a subject, every question type has its own way of asking what you know. Objective questions, such as multiple-choice or true/false, test your ability to recall, compare, and contrast information and to choose the right answer from among several choices. Subjective questions, usually essay questions, demand the same information recall but ask that you analyze the mind actions and thinking processes required, then organize, draft, and refine a written response. The following guidelines will help you choose the best answers to both types of questions.

TERMS

Objective questions Short-answer questions that test your ability to recall, compare, and contrast information and to choose the right answer from a limited number of choices.

Multiple-Choice Questions

Multiple-choice questions are the most popular type on standardized tests. The following strategies can help you answer these questions:

Subjective questions Essay questions that require you to express your answer in terms of your own personal knowledge and perspective.

Read the directions carefully. While most test items ask for a single correct answer, some give you the option of marking several choices that are correct.

First read each question thoroughly. Then look at the choices and try to answer the question.

Underline key words and phrases in the question. If the question is complicated, try to break it down into small sections that are easy to understand.

Pay special attention to qualifiers such as *only, except,* **etc.** For example, negative words in a question can confuse your understanding of what the question asks ("Which of the following is *not* . . .").

If you don't know the answer, eliminate those answers that you know or suspect are wrong. Your goal is to narrow down your choices. Here are some questions to ask:

- Is the choice accurate in its own terms? If there's an error in the choice—for example, a term that is incorrectly defined—the answer is wrong.
- Is the choice relevant? An answer may be accurate, but it may not relate to the essence of the question.
- Are there any qualifiers, such as *always, never, all, none,* or *every?* Qualifiers make it easy to find an exception that makes a choice incorrect. For example, the statement that "children *always* begin talking before the age of two" can be eliminated as an answer to the question, "When do children generally start to talk?"
- Do the choices give you any clues? Does a puzzling word remind you of a word you know? If you don't know a word, does any part of the word (prefix, suffix, or root) seem familiar to you?

"A little knowledge that acts is worth infinitely more than much knowledge that is idle."
KAHLIL GIBRAN

Look for patterns that may lead to the right answer, then use intelligent guessing. Test-taking experts have found patterns in multiple-choice questions that may help you get a better grade. Here is their advice:

- Consider the possibility that a choice that is more *general* than the others is the right answer.
- Look for a choice that has a middle value in a range (the range can be from small to large, from old to recent). This choice may be the right answer.
- Look for two choices with similar meanings. One of these answers is probably correct.

Make sure you read every word of every answer. Instructors have been known to include answers that are right except for a single word.

When questions are keyed to a long reading passage, read the questions first. This will help you focus on the information you need to answer the questions.

Here are some examples of the kinds of multiple-choice questions you might encounter in an Introduction to Environmental Science course (the correct answer follows each question):

1. Significant increases in carbon dioxide levels have resulted from
   a. photosynthesis by land plants and algae
   b. deforestation and burning fossil fuels
   c. the evaporation of seawater
   d. the greenhouse effect

   (Correct answer is B)

2. Human actions that might reasonably reduce the greenhouse effect include
   a. shifting to using more synthetic fuels
   b. maintaining food reserves
   c. building coastal dikes
   d. planting more trees

   (Correct answer is A)

3. Which of the following would function as an output control device for motor vehicles?
   a. improved motor efficiency
   b. rely on mass transit and bicycles
   c. use emission control devices
   d. tax new cars based on their efficiency

   (Correct answer is C)

True/False Questions

True/false questions test your knowledge of facts and concepts. Read them carefully to evaluate what they truly say. Try to take these questions at face value without searching for hidden meaning. If you're truly stumped, guess (unless you're penalized for wrong answers).

Look for qualifiers in true/false questions—such as *all, only, always, because, generally, usually,* and *sometimes*—that can turn a statement that would otherwise be true into one that is false, or vice versa. For example, "The grammar rule, 'I before E except after C,' is *always* true" is *false*, whereas "The grammar rule, 'I before E except after C,' is *usually* true," is *true*. The qualifier makes the difference. The box on the next page offers some examples of the kinds of true/false questions you might encounter in an Introduction to Psychology course (the correct answer follows each question).

Essay Questions

An essay question allows you to use writing to demonstrate your knowledge and express your views on a topic. Start by reading the questions and deciding which to tackle (sometimes there's a choice). Then focus on what each

> Are the following questions true or false?
>
> 1. Alcohol use is always related to increases in hostility, aggression, violence, and abusive behavior. (False)
> 2. Marijuana is harmless. (False)
> 3. Simply expecting a drug to produce an effect is often enough to produce the effect. (True)

question is asking, the mind actions you will have to use, and the writing directions. Read the question carefully, and do everything you are asked to do. Some essay questions may contain more than one part.

Watch for certain action verbs that can help you figure out what to do. Figure 8-2 explains some words commonly used in essay questions. Underline these words as you read any essay question and use them as a guide.

Next, budget your time and begin to plan. Create an informal outline or think link to map your ideas and indicate examples you plan to cite to support those ideas. Avoid spending too much time on introductions or flowery prose. Start with a thesis idea or statement that states in a broad way what your essay will say (see Chapter 7 for a discussion of thesis statements). As you continue to write your first paragraph, introduce the essay's points, which may be sub-ideas, causes and effects, or examples. Wrap it up with a concise conclusion.

Use clear, simple language in your essay. Support your ideas with examples, and look back at your outline to make sure you are covering everything. Try to write legibly. If your instructor can't read your ideas, it doesn't matter how good they are. If your handwriting is messy, try printing, skipping every other line, or writing on only one side of the paper.

Do your best to save time to reread and revise your essay after you finish getting your ideas down on paper. Look for ideas you left out and sentences that might confuse the reader. Check for mistakes in grammar, spelling, punctuation, and usage. No matter what subject you are writing about, having a command of these factors will make your work all the more complete and impressive.

Here are some examples of essay questions you might encounter in your Introduction to Physiology course. In each case, notice the action verbs from Figure 8-2.

> 1. Describe how different sense organs enable the body to obtain information about internal and external environments. Include specifics of taste, vision, hearing, touch, and smell.
> 2. What are the major glands of the endocrine system; what hormone(s) does each produce; and what are their affects on the body?

Analyze Break into parts and discuss each part separately.

Compare Explain similarities and differences.

Contrast Distinguish between items being compared by focusing on differences.

Criticize Evaluate the positive and negative effects of what is being discussed.

Define State the essential quality or meaning. Give the common idea.

Describe Visualize and give information that paints a complete picture.

Discuss Examine in a complete and detailed way, usually be connecting ideas to examples.

Enumerate/List/Identify Recall and specify items in the form of a list.

Explain Make the meaning of something clear, often by making analogies or giving examples.

Evaluate Give your opinion about the value or worth of something, usually by weighing positive and negative effects, and justify your conclusion.

Illustrate Supply examples.

Interpret Explain your personal view of facts and ideas and how they relate to one another.

Outline Organize and present the sub-ideas or main examples of an idea.

Prove Use evidence and argument to show that something is true, usually by showing cause and effect or giving examples that fit the idea to be proven.

Review Provide an overview of ideas, and establish their merits and features.

State Explain clearly, simply, and concisely, being sure that each word gives the image you want.

Summarize Give the important ideas in brief.

Trace Present a history of the way something developed, often by showing cause and effect.

Figure 8-2

Common action verbs on essay tests.

How can you learn from test mistakes?

The purpose of a test is to see how much you know, not merely to achieve a grade. The knowledge that comes from attending class and studying should allow you to correctly answer test questions. Knowledge also comes when you learn from your mistakes. If you don't learn from what you get wrong on a test, you are likely to repeat the same mistake again on another test and in life. Learn from test mistakes just as you learn from mistakes in your personal and business life.

Try to identify patterns in your mistakes by looking for:

- *Careless errors*—In your rush to complete the exam, did you misread the question or directions, blacken the wrong box, skip a question, or use illegible handwriting?

◆ *Conceptual or factual errors*—Did you misunderstand a concept or never learn it in the first place? Did you fail to master certain facts? Did you skip part of the assigned text or miss important classes in which ideas were covered?

You may want to rework the questions you got wrong. Based on the feedback from your instructor, try rewriting an essay, recalculating a math problem, or redoing the questions that follow a reading selection. As frustrating as they are, remember that mistakes show that you are human, and they can help you learn. If you see patterns of careless errors, promise yourself that next time you'll try to budget enough time to double-check your work. If you pick up conceptual and factual errors, rededicate yourself to better preparation.

When you fail a test, don't throw it away. First, take comfort in the fact that a lot of students have been in your shoes and that you are likely to improve your performance. Then recommit to the process by reviewing and analyzing your errors. Be sure you understand why you failed. You may want to ask for an explanation from your instructor. Finally, develop a plan to really learn the material if you didn't understand it in the first place.

སེམས་མ་ཡེངས་ཤིག

In Sanskrit, the written language of India and other Hindu countries, the characters above read *sem ma yeng chik*, meaning, "do not be distracted." This advice can refer to focus for a task or job at hand, the concentration required to critically think and talk through a problem, or the mental discipline of meditation.

Think of this concept as you strive to improve your listening and memory techniques. Focus on the task, the person, or the idea at hand. Try not to be distracted by other thoughts, other people's notions of what you should be doing, or any negative messages. Be present in the moment to truly hear and remember what is happening around you. Do not be distracted.

Chapter 8   Applications

Name _____ Date _____

KEY INTO YOUR LIFE
Opportunities to Apply What You Learn

 Optimum Listening Conditions

Describe a recent classroom situation in which you had an easy time listening to the instructor:

Where are you? _____

What is the instructor discussing? _____

Is it a straight lecture, or is there give-and-take between instructor and students? _____
_____

What is your state of mind? (List factors that might affect your ability to listen.) _____
_____
_____
_____

Are there any external barriers to communication? If yes, what are they, and how do they affect your concentration? _____
_____
_____
_____

Now describe a situation where you have found it more difficult to listen.

Where are you? _____

What is the instructor discussing? _____

Is it a straight lecture, directions for a lab, or is there give-and-take between instructor and students? _____
_____

What is your state of mind? (List factors that might affect your ability to listen.) _____
_____
_____
_____

Are there any external barriers to your hearing, physical or environmental? If yes, what are they, and how do they affect your concentration? _____
_____
_____
_____

Examine the two situations. Based on your descriptions, name three conditions that are crucial for you to listen effectively.

1. _____
2. _____
3. _____

What steps can you take to re-create these conditions in more difficult situations like the second one you described? _____
_____
_____
_____

8.2 Create a Mnemonic Device

Look back at all the memory principles examined in this chapter. Using what you learned about mnemonic devices, create a mnemonic that allows you to remember these memory principles quickly. You can create a mental picture or an acronym. If you are using a mental picture, describe it here; if you are using an acronym, write it and then indicate what each letter stands for.

_____
_____
_____
_____
_____

Think of other situations in which you used a mnemonic device to remember something. What was the device? How effective was it in helping you remember the information?

8.3 Learning From Your Mistakes

For this exercise, use an exam on which you made one or more mistakes.

- Why do you think you answered the question(s) incorrectly?

- Did any qualifying terms, such as *always, sometimes, never, often, occasionally, only, no,* and *not,* make the question(s) more difficult or confusing? What steps could you have taken to clarify the meaning?

- Did you try to guess the correct answer? If so, why do you think you made the wrong choice?

- Did you feel rushed? If you had had more time, do you think you would have gotten the right answer(s)? What could you have done to budget your time more effectively?

- If an essay question was a problem, what do you think went wrong? What will you do differently the next time you face an essay question on a test?

___

KEY TO SELF-EXPRESSION
Discovery Through Journal Writing

To record your thoughts, use a separate journal or the lined pages at the end of the chapter.

Write about how you feel about tests and how you generally perform when you take them.

As you walk into a room for a test, does your heart race or your mind go blank? Do you feel apprehensive? Does your performance on tests accurately reflect what you know or do your tests scores fall short of your knowledge? If there is a gap between your knowledge and your scores, why do you think this gap exists? What can you do to work through any test anxiety you have?

Journal

Name _____ Date _____

Journal

Name _____ Date _____

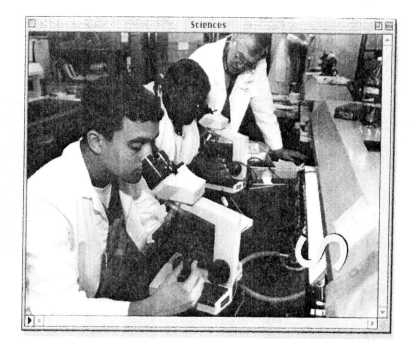

Working in the Laboratory

Safe Science in Action

Basically, there are two types of labs. One type, confirmatory, gives you the chance to better understand, or confirm, concepts you are studying in the classroom. You will do predetermined experiments that let you see for yourself evidence that confirms theories (also known as deductive). The second, original inquiry, gives you the opportunity to learn skills needed to set up an experiment to answer original questions. In either case, experiments done in undergraduate labs involve doing research by learning methods for collecting data to test theories, confirm hypotheses, or answer questions, raised by a scientist.

This chapter will help prepare you for your lab classes, and if you are a returning student who hasn't been in a lab class for some time, familiarize you with the basic do's and don'ts for safe labs. Inquiry-based lab science will be explored along with the practical considerations such as credit hours and equipment.

In this chapter, you will explore the answers to the following questions:

- What is inquiry-based research?
- What equipment will you need in the lab?
- How do credit hours translate into lab time?
- What can you do to boost your chances for success?
- Why is it important to practice safe lab science?
- What skills does lab work give you?

WHAT IS INQUIRY-BASED RESEARCH?

When the answers are not known you may take on the role of a researcher. Inquiry is just as it sounds, an inquiry, or exploration, into something you desire to know.

The *American Heritage College Dictionary* defines inquiry as follows:

1. The act of inquiring.
2. A question; a query.
3. A close examination of a matter in a search for information or truth.

"Twinkle, twinkle little star
I don't wonder what you are
For by spectroscopic ken I know that you are hydrogen"
ANONYMOUS

As the definition states, inquiry is about asking questions and carefully examining evidence that helps you answer the questions. This is usually done by forming a hypothesis and then deciding how to test your hypothesis.

A hypothesis is like an assumption, or a likely explanation to your question. If you are wondering about the affect of alcohol consumption during pregnancy, you might ask the question, "Do women who drink alcohol during pregnancy have infants that are small and poorly developed?" From this you develop a testable hypothesis:

Birth weight is lower among infants of alcoholic women than among infants of nonalcoholic women.

To test this directly you would have to have access to patient charts so you could look for alcoholic women who drank during their pregnancy and nonalcoholic women who didn't and then look for the weights of their infants and compare. Were the infants of women who drank smaller than those whose women didn't drink? In the lab you could devise an experiment to see if alcohol affects individual cell development.

Often the answers to your questions are not known. This gives you an opportunity, in the lab or in the field, to take on the role of the research scientist. You will learn how to:

1. Ask questions
2. Form ways to answer your questions, either experimentally or through observation.
3. Collect information and record it for later use or analysis.

Qualities of a Researcher

Labs, especially those based on inquiry, teach you many qualities of a research scientist. These qualities include creativity, patience, persistence.

Creativity. You have to find different ways to answer your questions, which requires looking at things from different viewpoints, understanding your own biases, and letting your mind wander; daydreaming can be an important aspect of the creative process.

Patience. You must possess, or learn to possess, a willingness to not have clear answers right away. Very little in science is black and white, yes or no. You need to tolerate some degree of ambiguity, or uncertainty. Research requires time, repeating your work, and meticulous attention to detail.

Persistence. Because the questions and answers are often not immediately clear, you must be able to continue searching, experimenting, and observing even when you are not getting results right away. Darwin, for instance, spent years analyzing his specimen collections from his journey on the H.M.S. *Beagle* before he formed his theory to explain evolution.

Levels of Inquiry

In the lab you will really begin to learn about science by practicing science. The key to success in the lab is learning to think for yourself. There are three levels of inquiry and going through them will help lead you toward learning to think for yourself.

You will start in middle school, high school, and often as an undergraduate, with the first level, or "recipe lab" science. This is when you are given exact instructions, like a recipe, on how to do the lab and what your results should look like. This is good for learning how to follow instructions, use the equipment, and compare your results to a known standard.

The next level is guided inquiry. At this level the instructor knows the answers to predetermined questions. The instructor helps you find the methods needed to answer the question. You begin to think for yourself and to make some choices about how to proceed in your experiments.

At the third level, the student does it all. You decide what questions to ask, you develop a hypothesis, and you choose and decide upon methods for answering those questions.

Inquiry, according to research scientist Sylvia Oliver, Ph.D., is what makes everybody a scientist. "From the time you wake up in the morning," she says, "until you go to sleep at night, you ask questions about the world around you." Inquiry is simply observation sparking questions that lead to assumptions and eventually to decisions on what steps to take to answer the question, or test the truth of your assumption. This process requires that you do not accept what you hear, read, or even see, as fact. You must always ask "why," and when you learn to do this, you will be on your way to becoming a creative scientist.

WHAT EQUIPMENT WILL YOU NEED IN THE LAB?

You probably already know the basic equipment needed for any lab:

Lab coat to protect your clothing

Safety eyeglasses to protect your eyes

Latex or nonlatex (if you are sensitive or allergic to latex) gloves

Lab books, notebooks to record data

Other equipment needed for labs will vary from course to course. Most will require an additional lab fee to cover the costs of materials. Some equipment you use will be very complex and technical, such as that used for electrophoresis, and others will be more simple, such as test tubes and agar plates.

Check the course syllabus to see what you need to buy. Usually, any required equipment will be available in the school's bookstore. Borrowing equipment is a possible way to cut costs, as long as the equipment is in good condition. If you can find students in your other classes who have previously taken the course, ask if you can borrow or buy their used equipment.

HOW DO COURSE CREDIT HOURS TRANSLATE INTO LAB TIME?

Knowing how many hours per week you will be in lab helps you plan your time. Usually the number of course credit hours you take equals the number of hours per week you spend in the classroom. If you attend a school that breaks the year into semesters, each credit means one hour per week spent in class. If you take a three-credit class, you spend three hours a week attending class. However, for a lab, the number of hours spent in the lab is usually at least doubled. For a two-credit lab course, you may spend four to six hours a week in the lab.

The reason you are given more time in the lab is that unlike lectures, where large amounts of information can be conveyed in a short time, lab work requires longer to see results. If you are coming into college directly out of high school, this is not a new concept for you. If, however, you are returning to school to pursue a second career, it may have been some time since you took a lab course. Labs require preparation time before each session, scheduled lab time, and time after the lab to write up your work.

What Can You Do to Boost Your Chances for Success?

Four keys to success in the lab are:

1. Preparation: Taking time to prepare before each lab.
2. Attendance: Showing up at every lab.
3. Writing: Taking the time during and immediately after the lab to write up your work. In other words, don't rely on your memory.
4. Curiosity: Learning to ask why.

Preparation

Instructors will tell you that preparation is the key to success in the lab. Why? Because you must know what you will be doing before you do it; think of it as rehearsing for a play, a concert, a speech, or any other kind of performance. Preparation is time well spent because it will save you time once you are in the lab, allowing time to actually perform the experiments instead of reading the instructions.

Another important outcome of good preparation is that you will get better results if you know beforehand what you are going to do and how you are going to do it; that's the rehearsal part. "Always be prepared" is a good motto for lab classes.

TO PREPARE:

- Read through the lab at least once.
- Picture the lab step-by-step all the way through to the results. This is a mental rehearsal, done just like a downhill skier, snow boarder, or any other competitor who visualizes and memorizes the course before they are on it.
- Write one or two pages about the lab and highlight any questions you may have. Make these "what would happen if" questions to discuss with your lab instructor (often a graduate student) during or after class.

Attendance

Labs are more difficult to make up than lectures, and you will want to make up any missed labs. If you know you will be unable to attend a lab session, you must contact the instructor beforehand, either by phone, e-mail, or in person, to let them know. When you talk to them, be sure to discuss how you can make up the lab. If a sudden emergency occurs, so that you cannot contact the instructor before the missed lab, contact them as soon as you are able. It shouldn't be necessary to make long explanations. Simply tell them in a sentence or two that you will not be at the lab and why. If your reasons are legitimate, there is no need to apologize, plead, or complain in any way. Simply state the facts.

Writing

Although most of your lab time will be spent on preparation and doing the actual lab work, you are required to write up your work. The purpose for this is to show others what you did. This is a basic tenet of the research process: leaving a trail that can be followed and repeated. This is how research findings are verified. Repetition of studies and research methods is done to examine the appropriateness and accuracy of a study's results.

There are many formats for lab write-ups, and it is certain that your instructors will tell you what is expected in their courses. This means that from lab course to lab course you may change write-up formats, but essentially the information you include will be the same. Figure 9-1 is an example of a lab write-up for a microbiology course; look for format and required components.

Curiosity

You are working in the microbiology lab observing bacteria cells you cultured the day before. They appear healthy and thriving, except that on closer inspection you notice an area without cell growth. You try to remember preparing the culture, and wonder if you accidentally contaminated the plate. This starts a string of questions beginning with, "Is something contaminating the plate that inhibited bacterial growth?" You are familiar with serendipity, and although you consider it unlikely that you will make a major discovery in sophomore microbiology class, you are curious about the phenomenon you are observing.

Curiosity is the desire to make further inquiry. It means being inquisitive about new observations, new situations, and new questions. It also means not taking everything you see, or are told, as fact. Questioning is the outcome of curiosity, and questions are at the root of hypothesis and, therefore, research science. How did anyone decide to check out if the earth was really flat? They considered their observation of the curved horizon and then asked: "What if the earth could be curved, or even round?" To test this idea, they hypothesized that the earth was round and proposed sailing to the edge to test their hypothesis. This took courage as well as curiosity.

WHY IS IT IMPORTANT TO PRACTICE SAFE LAB SCIENCE?

No doubt in high school you were thoroughly drilled about lab safety, but again, if you haven't been in school for a while, a refresher helps. Instructors are always passionate about lab safety. Why? Because there are many things that can go wrong, leading to injury, equipment damage, or a failed experiment (which equals wasted time).

Laboratories are full of expensive and delicate equipment, and potential hazards: electronics, glassware, caustic chemicals, flammable chemicals and gases, sensitive measurement instruments, bacteria and viruses, automated machinery, potentially slippery floors, and most of all, human bodies. Squirting liquids or moving objects can easily find their way into eyes. Toxic

Figure 9-1 Microbiology lab write-up.

Bacterial Conjugation　　　　F-duction in "E. Coli"　　　　7/1/99

Following protocol, the F+/F− mix was placed in the 37 degree C incubation at 1:44 PM.

3:45 PM: Began streaking plates as per protocol on the control plates. All the streaks were completed and placed in the 37 degree C incubator at 3:59 PM.

MacConkey agar used (Beta − gal+ colonies will be red).

To be incubated overnight.

July 2/99 1:00 PM

	Plate	Presence of colonies	Colony color	Lac+ Lac−	Strep resistance Strep sensitive
Controls with Streptomycin	F+ (male)	none			Sensitive
	F− (female)	Yes	cream	Lac−	Resistant
Controls without Streptomycin	F+	Yes	red	Lac+	
	F−	Yes	cream	Lac−	
Mated		Yes	red &	Lac+	Resistant
			cream	Lac−	Resistant

Conclusions: The mated plate produced the colonies we expected. The cream-colored colonies would be F− bacteria that were not conjugated and due to their Strep resistance can still grow. The red colonies were evidence of successful conjugation: F− bacteria with Strep resistance having received Lac+ from the F+ strain. These results assume that the information given on F+, or F−, was accurate.

---

fumes from burning or evaporation can find their way into nostrils and mouths, leading to the tender bronchus, lung tissue, and alveoli.

The list could go on indefinitely, but as you know, everything you do has risk. The important thing is to minimize the risk to yourself and to others by following the lab rules.

BASIC LAB RULES

- Know where the chemical shower and eye wash are located and how to use them.
- Have all the equipment ready before you start the lab.

- Follow the instructions.
- Understand the fire safety rules.
- Pay attention to what is going on around you.

WHAT SKILLS DOES LAB WORK GIVE YOU?

Lab work gives you the chance to practice and improve on many skills, not just the skills needed to carry out the experiment. You will learn how to use the equipment and set up experiments; you will learn how to work as part of a team, think creatively and analytically, be precise and write objectively; and you will learn how to communicate effectively.

Communication Skills

Can you write a clear, concise, and persuasive cover letter for a job interview, for a grant proposal, or for a scholarship application? Can you articulate the nature of your pursuit in science to people who know nothing about your subject, as well as to experts in your field? Communication skills are an integral part of science, and you will practice them in the lab setting during your classes when you work with your lab partners, talk to your professor, write up your lab findings, and present lab results to your peers.

Communication is at the top of every employer's list of desired employee qualities. The workplace is busy and complex requiring you to interact with a variety of people in a variety of professions. You must be able to speak and write clearly and coherently about the work you are doing and you must do so to many different people. For instance, you may be required to justify the importance of your work to your supervisor or present your work to a congressional panel or to public, private, or professional groups. Your skills in communicating may make the difference between gaining or losing funding for your work. Procuring funding for your work will always be a challenge and having optimal communication skills will benefit you in this area.

Teamwork Skills

Your lab classes will be done in small groups, or with a partner. A good team worker is something all employers are looking for, not only in research, but in all areas of science. Spending time in meetings to share results, strategize, make decisions, and plan budgets is something everyone does in the sciences. Writing scientific papers or preparing presentations also involves teamwork. Your science labs offer you an excellent opportunity to practice team skills.

Research is often done in teams with one person designated as the principal investigator, or leader, of the team. In universities your professors may be doing their own research, in which case he or she would be the principal investigator. Graduate students often make up the research team members. To gain more experience in research, and to practice team skills, you may be able to volunteer, or even find a work-study job in a professor's lab.

Critical-Thinking Skills

In the lab, especially as an undergraduate, you will most often be following a set of instructions to reach a specific result (confirmatory lab). Use your natural curiosity and inquisitiveness to practice critical thinking skills in these labs. Here are steps you can take to hone your thinking skills and add challenge to your labs:

- Read all lab instructions.
- Picture yourself doing each step of the lab and checking your results.
- Write several pages describing the lab.
- Write down questions that come to mind while studying.

For instance, in a histology lab you would be studying different types of tissue cells. Preparing for the lab may be very straightforward, but that does not keep you from asking questions that begin with the query, "What would happen if . . ." These are the types of questions research is based on. You can discuss your questions with the lab instructor or do research in the library on your own to begin finding answers to your questions. This process of asking questions and seeking answers is fundamental to improving your critical-thinking skills, and thus to becoming a better scientist.

Technical Skills

Technology is changing every day, and working in a lab is an opportunity to learn how new technology can be used to better understand old questions. Technology can also be used to begin asking new questions not considered in the past because the means for answering them did not exist.

Genetics and biotechnology is an example. The field of genetics has been around for over a century, beginning with recognizing inherited phenomena observable with the human eye (phenotype). Gregor Mendel did the first groundbreaking work in genetics by observing peas and their variations through breeding. Although Mendel wondered what caused these differences to occur, it wasn't until technology provided the means to discover them that genes controlling the visible pea traits were known.

As technology has helped us gain new knowledge in genetics, research such as the international Human Genome Project is well on the way to mapping the entire human genome. Learning how genes control disease is leading to advances in gene therapy research and the hope that someday genetic diseases may be treated or prevented.

A Summary for Lab Success

BE PREPARED BEFORE YOU ATTEND LAB:

- Read and reread the instructions.
- Picture yourself doing each step.
- Picture the results.

- Write one to two pages about the lab.
- Ask further questions.
- Be familiar with the purpose and goal of the lab.

FOLLOW INSTRUCTIONS:

- Follow lab instructions carefully.
- During lab, write down everything you do.
- Don't prejudge; nothing is insignificant.

LEARN SKILLS:

- Practice technical skills.
- Practice communication in the lab.
- Work in teams by cooperating with others.
- Sharpen math skills used in data analysis, volume conversions, measurement.
- Use precision and observation skills.
- Record events objectively and accurately.
  And don't forget: attendance is crucial to your success.

Docendo Discimus

This Latin phrase means we learn by teaching. As a student you may think that you are doing all the learning and your instructors all the teaching. But, you teach when helping other students in laboratories and in study groups and you teach your instructors through inquiring questions, thoughtful answers, and interesting life experiences. The concept that everyone learns as they teach emphasizes the cyclical and ongoing nature of education.

Chapter 9   Applications

Name _____ Date _____

KEY INTO YOUR LIFE
Opportunities to Apply What You Learn

 Inquiry in Action

Read the following research situation from *Science News*.[1] (Reprinted with permission from *Science Service*, the weekly newsmagazine of science, copyright 1998.)

WHY FLORIDA'S CORMORANTS LOOK DRUNK

Veterinarians may at last have found the cause of a mysterious and sometimes fatal disorder that turns big birds into fuddled wrecks with an odd stance, a goofy walk, and a tendency to shake their heads.

"People describe them as drunk," says veterinarian Christine Kreuder of Peninsula Equine Veterinarian Hospital in Menlo Park, California. However, the real problem for double-crested cormorants along Florida's Gulf Coast is probably due to toxic algae blooms, Kreuder told a joint meeting of the American Associations of Wildlife and Zoo Veterinarians in Omaha, Neb., on October 2.

Afflicted birds have been turning up for some 20 years. The wildlife clinic on Sanibel Island, Fla., where Kreuder used to work, received 388 sick cormorants in 1995 and 1996, but Kreuder failed to find a viral or bacterial cause.

Pathologist Gregory D. Bossart of the University of Miami is developing a test for a toxin produced by algae. When Kreuder sent him tissue samples in 1997, Bossart found that toxin. He also detected it in Common Mures that got sick in Monterey Bay, Calif., David A. Jessup of the California Department of Fish and Game in Santa Cruz reported at the meeting. Kreuder points out that the findings in the two species provide the first direct evidence for toxic algae effects on wild birds.

1. Identify the problem, and write a research question that could have been used to direct the research done in this situation. _____
_____
_____

2. What lab experiments could be conducted to answer your question? What lab equipment would you need? _____
_____
_____

3. What field studies could be conducted and what equipment would you need? _____
_____
_____
_____

4. What further experiments need to be done by the veterinarians, the pathologist, and the Fish and Game Department? _____
_____
_____

5. What is the logic that led to the discovery of the toxin? _____
   a. _____
   b. _____
   c. _____

6. What important theory, or background knowledge, did the veterinarian Kreuder need to have to make the decision to send the tissue samples for testing? _____
_____
_____

 ## 9.2 Using World Wide Web Science Sites

Using the Internet, go to one of the following sites and, using information found there, write several questions of your own that could be developed into experiments. Write them in the space on the following page, and be sure to include which site you visited.

   http://uga.edu/~protozoa
   www.science.news.com
   www.discover.com
   www.utdallas.edu/research/issues
   http://pao.gsfc.nasa.gov/gsfc/newsroom/flash/flash.htm
   www.4woman.org
   www.researchpaper.com
   www.nsf.org

www.nursingworld.org

1. _____
2. _____
3. _____
4. _____
5. _____
Site: _____

KEY TO SELF-EXPRESSION
Discovery Through Journal Writing

To record your thoughts, use a separate journal or the lined page at the end of the chapter.

Writing Questions

Think of at least one science topic that interests you. Write one or two sentences that sums up a problem in that topic area. Now, using that problem, write questions you could use to design experiments to answer them. Just brainstorm and let your mind "think up" whatever problems and questions you can.

"Contrary to what I once thought, scientific progress did not consist simply in observing, in accurately formulating experimental facts and drawing up a theory from them, it began with the invention of a possible world, or a fragment thereof, which was then compared by experimentation with the real world. And it was this constant dialogue between imagination and experiment that allowed one to form an increasingly fine-grained conception of what is called reality."

FRANÇOIS JACOB

CHAPTER 9 Working in the Laboratory

Journal

Name _____ Date _____

10
Relating to Others

Appreciating Your Diverse World

The greater part of your waking life involves interaction with people—family and friends, peers, fellow students, co-workers, instructors, and many more. When you put energy into your relationships and open the lines of communication, you can receive much in return. Having a strong network of relationships can help you grow and progress toward your goals.

This chapter will explore the issues of diversity that can hinder or help how you perceive others and relate to them. You will also explore communication styles, personal relationships, and the roles you can play in groups and on teams. Finally, you will read about various kinds of conflict and criticism, examining how to handle them so they benefit you instead of setting you back.

In this chapter, you will explore answers to the following questions:

- How can you understand and accept others?
- How can you express yourself effectively?
- How do your personal relationships define you?
- How can you handle conflict and criticism?
- What role do you play in groups?

HOW CAN YOU UNDERSTAND AND ACCEPT OTHERS?

Human interaction is an essential element of life. In a diverse world, many people are different from what you are familiar with and perceive as "normal." In this section you will explore diversity in your world, the positive effects of accepting diversity, and how to overcome barriers to understanding. The first requirements for dealing with differences are an open mind and a willingness to learn.

Diversity in Your World

For centuries, travel to different countries was seen as part of a complete education. Edith Wharton, a nineteenth-century author, wrote a story called "False Dawn" in which a character named Mr. Raycie recommends travel to his son: "A young man, in my opinion, before setting up for himself, must see the world; form his taste; fortify his judgment. He must study the most famous monuments, examine the organization of foreign societies, and the habits and customs of those older civilizations . . . I believe he will be able to learn much."[1] When cultures were so separated, learning about differences was best accomplished through travel.

Today, although traveling is still a valuable way to learn, different places and people often come to you. More and more, diversity is part of your community, on your television, on your Internet browser, at your school, in your workplace, and in your family. It used to be that most people lived in societies with others who seemed very similar to them. Now, differences are often woven into everyday life.

You may encounter examples of diversity like these:

- Communities with people from different stages of life
- Co-workers who represent a variety of ethnic origins
- Classmates who speak a number of different languages
- Social situations featuring people from various cultures, religions, and sexual orientations
- Individuals who marry a person or adopt a child from a different racial or religious background

TERMS

Diversity
The variety that occurs in every aspect of humanity, involving both visible and invisible characteristics.

- Diverse restaurants, services, and businesses in the community
- Neighborhoods with immigrants from a variety of social/class backgrounds
- Different lifestyles as reflected in books, magazines and newspapers, television, movies, music, the Internet, and other forms of popular culture
- People in the workplace who have a variety of disabilities—some more obvious than others

Each person has a choice about how to relate to others—or *whether* to relate to them. No one can force you to interact with any other person, or to adopt a particular attitude as being "right." Considering two important responsibilities may help you sort through your options.

Your responsibility to yourself lies in being true to yourself, in taking time to think through your reactions to other people. When you evaluate your thoughts, try to also consider their source: Have you heard these ideas from other people or organizations or the media? Do you agree with them, or does a different approach feel better to you? Through critical thinking you can arrive at decisions about which you feel comfortable and confident.

Your responsibility to others lies in treating people with tolerance and respect. No one will like everyone he or she meets, but acknowledging that others deserve respect and have a right to their opinions will build bridges of communication. The more people accept one another, the more all kinds of relationships will be able to thrive.

The Positive Effects of Diversity

More than just "a nice thing to do," accepting diversity has very real benefits to people in all kinds of relationships. Acceptance and respect form the basis for any successful interaction. As more situations bring diverse people into relationships, communication will become more and more dependent upon acceptance and mutual understanding.

Consider how positive relationships with diverse people may contribute to success. Relationships among family, friends, and neighbors affect personal life. Relationships among students, instructors, and other school personnel affect student life. Relationships among co-workers, supervisors, and customers/clients affect work life. Understanding and communication in these relationships can bring positive effects such as satisfying relationships, achievement, and progress. Failure to understand and communicate well can have negative effects.

"Minds are like parachutes. They only function when they are open."

SIR JAMES DEWAR

For example, examine the potential effects of reactions to diversity in the following situations. Although each of these situations focuses on the reaction of only one person, it's important to note that both parties need to work together to establish mutual trust and openness.

A male Hispanic employee has a female African-American supervisor. If the employee believes negative stereotypes about women or African-American people and resists taking directions from the supervisor, he may lose his job or be viewed as a liability. On the other hand, if the employee can respect the supervisor's authority and consider any different methods or ideas she has, their relationship is more likely to become supportive and productive. He may then be more likely to feel comfortable, perform well, and move up at work.

A learning-disabled student has an Asian instructor. If the student assumes that Asian people are superior and intimidating, letting that opinion lead her to resist the advice and directions her instructor gives her, she may do poorly in the class or drop the course. On the other hand, if the student stays open to what the instructor has to offer, the instructor may feel respected and may be more encouraging. The student may then be more likely to pay attention in class, work hard, and advance in her education.

A Caucasian man has a sister who adopts a biracial child. If the man cuts off contact with his sister because he fears racial differences and doesn't approve of racial mixing, he may deny himself her support and create a rift in the family. On the other hand, if the man can accept the new family member and respect his sister's choice, she may feel more supported and continue to support him in turn. The situation may help to build a close and rewarding family relationship.

Accepting others isn't always easy, and it's common to let perceptions about people block your ability to communicate. Following are some barriers that can hinder your ability to accept and understand others, and suggestions for how to overcome them.

Barriers to Understanding

You deserve to feel positive about who you are, where you come from, what you believe, and the others with whom you identify. However, problems arise when people use the power of group identity to put others down or cut themselves off from others. Table 10-1 shows how an open-minded approach can differ from an approach that is characterized by barriers.

Stereotypes, prejudice, discrimination, and fear of differences can form barriers to communication.

Stereotypes

TERMS

Stereotype
A standardized mental picture that represents an oversimplified opinion or uncritical judgment.

As you learned in Chapter 5, an assumption is an idea that you accept without looking for proof. A stereotype occurs when an assumption is made about a person or group of people based on one or more characteristics. You may have heard stereotypical assumptions such as these: "Women are too emotional for business." "African-Americans can run fast and sing well." "Hispanics are Catholic and have tons of kids." "White people are cold and power-hungry." "Gay people sleep around." "Learning-disabled people can't hold down jobs." "Older people can't learn new things." Stereotypes are as common as they are destructive.

Why might people stereotype? The list below offers a few reasons.

Table 10-1 A closed-minded approach vs. an open-minded approach.

YOUR ROLE	SITUATION	CLOSED-MINDED APPROACH	OPEN-MINDED APPROACH
Team member on the job	A co-worker from India observes a Hindu religious ritual at lunchtime.	You stare at the religious ritual, thinking it weird. You feel that this co-worker should just blend in and act like everyone else.	Your observe the ritual, respecting how the person expresses religious beliefs. You look up Hindu religion in your spare time to learn more.
Fellow student	For an assignment, you are paired up with a student old enough to be your mother.	You figure that the student will be closed off to the modern world. You think that she might also act like a parent and preach to you about how to do the assignment.	You avoid thinking that this student will act like your parents and get to know her as an individual. You stay open to what you can learn from her experiences and knowledge.
Friend	You are invited to dinner at a friend's house for the first time. When he introduces you to his partner, you realize that he is gay.	You are turned off by the idea of two men in a relationship and by gay culture in general. You are uncomfortable and make an excuse to leave early. You avoid your friend from then on.	You have dinner with your friend and his partner. You learn that they have a committed, supportive relationship. You take the opportunity to learn more about who they are and what their lives are like.
Employee	Your new boss is Japanese-American, hired from a competing company.	You assume that your new boss is very hard-working, expecting unrealistic things from you and your co-workers. You assume she doesn't take time to socialize.	Your rein in your assumptions, knowing they are based on stereotypes, and approach your new boss with an open mind.

People seek patterns and logic. Trying to make sense of a complex world is part of human nature. People often try to find order by using the labels and categories that stereotypes provide.

Stereotyping is quick and easy. Making an assumption about a person from observing an external characteristic is easier than working to know a person as a unique individual. Labeling a group of people based on a characteristic they seem to have in common takes less time and energy than exploring the differences and unique qualities within the group.

Movies, magazines, and other media encourage stereotyping. The more people see stereotypical images—the unintelligent blonde, the funny overweight person, the evil white businessman—the easier it is to believe that such stereotypes are universal.

The ease of stereotypes comes at a high price. First and foremost, stereotypes can perpetuate harmful generalizations and falsehoods about others. These false ideas can promote discrimination. For example, if an employer believes that Vietnamese people cannot speak English well, he might not even bother to interview them. Secondly, stereotypes also communicate the message that you don't care about or respect others enough to discover who they really are. This may encourage others to stereotype you in return. Others may not give you a chance if they feel that you haven't given them a chance.

Addressing stereotypes. Recall from the critical-thinking material in Chapter 5 the questions you can ask about an assumption in order to examine its validity. Apply these questions to stereotypes:

1. In what cases is this stereotype true, if ever? In what cases is it not true?
2. Has stereotyping others benefited me? Has it hurt me? In what ways?
3. If someone taught me this stereotype, why? Did that person think it over or just accept it?
4. What harm could be done by always accepting this stereotype as true?

Using these steps, think through the stereotypes you assume are true. When you hear someone else use a stereotype and you know some information that disproves it, volunteer that information. Encourage others to think through stereotypes and to reject them if they don't hold up under examination.

Give others the benefit of the doubt. Thinking beyond stereotypes is an important step toward more open lines of communication.

Prejudice

Prejudice
A preconceived judgment or opinion, formed without just grounds or sufficient knowledge.

Prejudice occurs when people "prejudge," meaning that they make a judgment before they have sufficient knowledge upon which to base that judgment. People often form prejudiced opinions on the basis of a particular characteristic—gender, race, culture, abilities, sexual orientation, religion, and so on. You may be familiar with the labels for particular kinds of prejudice, such as *racism* (prejudice based on race) or *ageism* (prejudice based on age). Any group can be subjected to prejudice, although certain groups have more often been on the receiving end of such closed-minded attitudes. Prejudice can lead people to disrespect, harass, and put down others. In some cases, prejudice may lead to unrealistic expectations of others that aren't necessarily negative, such as if someone were to judge all Jewish people as excelling in business.

Prejudice can have one or more causes. Some common causes include the following.

People experience the world through the lens of their own particular identity. You grow up in a particular culture and family and learn their attitudes. When you encounter different ideas and ways of life, you may react by categorizing them. You may also react with ethnocentrism—the idea that your group is better than anyone else's.

When people get hurt, they may dislike or blame anyone who seems similar to the person who hurt them. Judging others based on a bad experience is human, especially when a particular characteristic raises strong emotions.

Jealousy and fear of personal failure can lead a person to want to put others down. When people are feeling insecure about their own abilities, they might find it easier to devalue the abilities of others rather than to take risks and try harder themselves.

The many faces of prejudice often appear on college campuses. A student may not want to work with an in-class group that contains people of another race. Campus clubs may tend to limit their membership to a particular group and exclude others. Religious groups may devalue the beliefs of other religions. Groups that gather based on a common characteristic might be harassed by others. Women or men may find that instructors or fellow students judge their abilities and attitudes based on their gender. All of these attitudes severely block attempts at mutual understanding.

Addressing prejudice. Being critical of people who are different cuts you off from all kinds of perspectives and people that can enhance the quality of your life. Critical thinking is your key to changing prejudicial attitudes. For example, suppose you find yourself thinking that a certain student in your class isn't the type of person you want to get to know. Ask yourself: Where did I get this attitude? Am I accepting someone else's judgment? Am I making judgments based on how this person looks or speaks or behaves? How does having this attitude affect me or others? If you see that your attitude needs to change, have the courage to activate that change by considering the person with an open, accepting mind.

Another tactic, and often an extremely difficult one, is to confront people you know when they display a prejudicial attitude. It can be hard to stand up to someone and risk a relationship or, if the person is your employer, even a job. On the other hand, your silence may imply that you agree. Evaluate the situation and decide what choice is most suitable and is true to your values. Ask yourself if you can associate with a person if he or she thinks or behaves in a way that you do not respect.

You have a range of choices when deciding whether to reveal your feelings about someone's behavior. You can decide not to address it at all. You may drop a humorous hint and hope that you make your point. You may make a small comment to "test the waters" and see how the person reacts, hoping that later you can have a more complete discussion about it. Whatever you do, express your opinion respectfully. Perhaps the other person will take that chance to rethink the attitude; perhaps he or she will not. Either way, you have taken an important stand.

Discrimination

Discrimination occurs when people deny others opportunities because of their perceived differences. Prejudice often accompanies discrimination. Discrimination can mean being denied jobs or advancement, equal educational opportunities, equal housing opportunities, services, or access to events, people, rights, privileges, or commodities.

Discrimination happens in all kinds of situations, revolving around gender, language, race, culture, and other factors. A 32-year-old married woman may not get a job because the interviewer assumes that she will become pregnant. A Russian person may be fired from a restaurant job because his English is heavily accented. Sheryl McCarthy, an African-American columnist for *New York Newsday*, sees it on the street. "Nothing is quite so basic and clear as having a cab go right past your furiously waving body and pick up the white person next to you," she says in her book, *Why Are the Heroes Always White?*[2] "Sometimes you can debate whether racism was the motivating factor in an act; here there is no doubt whatsoever." Even so-called majority populations may now experience the power of discrimination. For example, a qualified white man may be passed up for a promotion in favor of a female or minority employee.

The disabled are often targets of discrimination because people may believe that they are depressed and incapacitated. John Hockenberry, a wheelchair-using paraplegic who travels the world in his work as an award-winning journalist, challenges the idea that disabled people lead lives of unproductive misery. "My body may have been capable of less, but virtually all of what it could do was suddenly charged with meaning. This feeling was the hardest to translate to the outside, where people wanted to believe that I must have to paint things in this way to keep from killing myself," he says in his memoir *Moving Violations*.[3]

Obesity can invite discrimination as well. People who are overweight may have trouble winning jobs or moving up at work. Even shopping for clothing can present limited options. Only in recent years have certain brand-name designers begun to create clothing in women's sizes above 12, and many designers still discriminate.

Addressing discrimination. U. S. federal law states that it is unlawful for you to be denied an education, work, or the chance to apply for work, housing, or basic rights based on your race, creed, color, age, gender, national or ethnic origin, religion, marital status, potential or actual pregnancy, or potential or actual illness or disability (unless the illness or disability prevents you from performing required tasks, and unless accommodations for the disability are not possible). Unfortunately, the law is frequently broken, with the result that incidents go unnoticed. Many times people don't report incidents, fearing trouble from those they accuse. Sometimes people don't even notice their attitudes seeping through, such as in an interview situation.

First and foremost, be responsible for your own behavior. Never knowingly participate in or encourage discrimination. When you act on prejudicial attitudes by discriminating against someone, the barrier to communication this discrimination causes hurts you as well as anyone else involved. A person who feels denied and shut out may be likely to do the same to you, and may even encourage others to do so.

Second, if you witness a discriminatory act or feel that you have been discriminated against, decide whether you want to approach an authority about it. You may want to begin by talking to the person who can most directly affect the situation—an instructor, your supervisor, a housing authority. Don't assume that people know when they hurt or offend someone. For example, if you have a disability and you find that accommodations haven't been made for

REAL WORLD PERSPECTIVE

How can I adjust to diversity?

Carrie Nelson, University of Guadalajara—Centro de Estudios para Extranjeros

Once I made the decision to study Spanish, the idea of studying abroad in a Spanish-speaking country was simultaneous. I have always been fascinated with learning about other cultures and being introduced to new ideas. Once I started filling out forms for the school in Mexico and buying my plane ticket, however, I began to question my decision. I wondered if I could really handle this big of a change in my life. But in my heart I knew that I had made the right decision.

Now that I am actually studying in Guadalajara, Mexico, I cannot imagine how I could have been nervous. This is one of the best decisions I have ever made, although I have had to adjust to living here. In some aspects the Mexican and American cultures are very similar, but in other regards they can be very different. For example, in the United States the customer is considered the most important person and expects to be served very quickly. In Mexico the customer is important, but the way of life is much more relaxed. Therefore, if a clerk gets a personal phone call, they will most likely talk to that person whether there is a customer waiting or not.

Every day I experience something new or begin to better understand the differences in our cultures. By learning about other people, I am learning more about myself and the beauty and complexity of people. Still, I would like to deepen my relationship with the people of Guadalajara. I realize I am just scratching the surface of this experience. Do you have any suggestions?

Tan Pham, Gonzaga University—Spokane, Washington

In 1981, I escaped from Vietnam after having tried unsuccessfully 13 different times. Like your experience in Guadalajara, I found being in a new country both wonderfully enriching and overwhelming at the same time.

When I first arrived in the United States, I was amazed at the way the people value time. In Vietnam, as in Guadalajara, things are much slower. But here, time is seen as money. I learned it was very important to arrive at my appointments at the exact time they were scheduled. I was also amazed at the technology: the computers and the transportation.

When you visit another country, there is so much to learn. My recommendation is that you stay open to the experience. Let go of your past, and fully embrace the new life you have accepted. Remember not to isolate yourself. Talk to the people. Avoid socializing just with people from your own country. Try to immerse yourself fully in the new culture. Also, ask for help. Most people are very sincere in their desire to help you. Each land has unique opportunities. Take advantage of them as you explore your new world.

you at school, speak up. Meet with an advisor to discuss your needs for transport, equipment, or a particular schedule.

If you don't find satisfaction and change at that level, try the next level of authority (an administrator, your supervisor's boss, a government official). If that doesn't produce results, you can take legal action, although legal struggles can take a lot of time and drain a great deal of money out of your pocket. At each decision stage, weigh all the positive and negative effects and evaluate whether the action is feasible for you. Although keeping quiet may not bring change, you may not be able to act right away. In the long run, if you are able to stand up for what you believe, your actions may be worthwhile.

Fear of Differences

It's human instinct to fear the unknown. Many people stop long before they actually explore anything unfamiliar. They allow their fear to prevent them from finding out anything about what's outside their known world. As cozy as that world can be, it also can be limiting, cutting off communication from people who could enrich that world in many different ways.

The fear of differences has many effects. A young person who fears the elderly may avoid visiting a grandparent in a nursing home. A person of one religion might reject friendships with those of other religions out of a fear of different religious beliefs. Someone in a relationship may fear the commitment of marriage. A person might turn down an offer to buy a house in a neighborhood that is populated with people from a different ethnic group. In each case, the person may deny him or herself a chance to learn a new perspective, communicate with new individuals, and grow from new experiences.

Address fear of differences. Diversity doesn't mean that you have to feel comfortable with everyone or agree with what everyone else believes. The fear of differences, though, can keep you from discovering anything outside your own world. Challenge yourself by looking for opportunities to expose yourself to differences. Today's world increasingly presents such opportunities. You can choose a study partner in class who has a different ethnic background. You can expand your knowledge with books or magazines. You can visit a museum or part of town that introduces a culture new to you. You can attend an unfamiliar religious service with a friend. Gradually broaden your horizons and consider new ideas.

If you think others are uncomfortable with differences, encourage them to work through their discomfort. Explain the difference so that it doesn't seem so mysterious. Offer to help them learn more in a setting that isn't threatening. Bring your message of the positive effects of diversity to others.

Accepting and Dealing With Differences

Successful interaction with the people around you benefits everyone. The success of that exchange depends upon your ability to accept differences. How open can you be? Your choices range from rejecting all differences to freely celebrating them, with a range of possibilities in between. Ask yourself important questions about what course of action you want to take. Realize that the opinions of family, friends, the media, and any group with which you identify may sometimes lead you into perspectives and actions that you haven't thought through completely. Do your best to sort through outside opinions and make a choice that feels right.

At the forefront of the list of ways to deal with differences is mutual respect. Respect for yourself and others is essential. Admitting that other people's cultures, behaviors, races, religions, appearances, and ideas deserve as much respect as your own promotes communication and learning.

> "I have a dream that one day on the red hills of Georgia the sons of former slaves and the sons of former slave owners will be able to sit down together at the table of brotherhood."
> MARTIN LUTHER KING, JR.

What else can you do to accept and deal with differences?

- **Avoid judgments based on external characteristics.** These include skin color, weight, facial features, or gender.

Cultivate relationships with people of different cultures, races, perspectives, and ages. Find out how other people live and think, and see what you can learn from them.

Educate yourself and others. "We can empower ourselves to end racism through massive education," say Tamara Trotter and Joycelyn Allen in *Talking Justice: 602 Ways to Build and Promote Racial Harmony*.[4] "Take advantage of books and people to teach you about other cultures. Empowerment comes through education. If you remain ignorant and blind to the critical issues of race and humanity, you will have no power to influence positive change." Read about other cultures and people.

Be sensitive to the particular needs of others at school and on the job. Think critically about their situations. Try to put yourself in their place by asking yourself questions about what you would feel and do if you were in a similar situation.

Work to listen to people whose perspectives clash with or challenge your own. Acknowledge that everyone has a right to his or her opinion, whether or not you agree with it.

- **Look for common ground**—parenting, classes, personal challenges, interests.

Help other people, no matter how different they may be. Sheryl McCarthy writes about an African-American man who, in the midst of the 1992 Los Angeles riots, saw a man being beaten and helped him to safety. "When asked why he risked grievous harm to save an Asian man he didn't even know, Williams said, 'Because if I'm not there to help someone else, when the mob comes for me, will there be someone there to save me?'"[5] Continue the cycle of kindness.

Explore your own background, beliefs, and identity. Share what you learn with others.

Cultivate your own personal diversity. You may be one of the growing population of people who have two, three, or ten different cultures in your background. Perhaps your father is Native American and Filipino and Scottish and your mother is Creole (French, Spanish, and African-American). Respect and explore your heritage. Even if you identify only with one group or culture, there are many different sides of you.

- **Take responsibility for making changes instead of pointing the finger at someone else.** Avoid blaming problems in your life on certain groups of people.

Learn from the atrocities of history like slavery and the Holocaust. Cherish the level of freedom you have and seek continual improvement at home and elsewhere in the world.

Teach your children about other cultures. Impress upon them the importance of appreciating differences while accepting that all people have equal rights.

Recognize that people everywhere have the same basic needs. Everyone loves, thinks, hurts, hopes, fears, and plans. People are united through their essential humanity.

Expressing your ideas clearly and interpreting what others believe are crucial keys to communicating within a diverse world. The following section examines how you can communicate most effectively with the people around you.

How can you express yourself effectively?

The only way for people to know each other's needs is to communicate as clearly and directly as possible. Successful communication promotes successful school, work, and personal relationships. Exploring communication styles, addressing communication problems, and using specific success strategies will help you express yourself effectively.

Adjusting to Communication Styles

Communication is an exchange between two or more people. The speaker's goal is for the listener (or listeners) to receive the message exactly as the speaker intended. Different people, however, have different styles of communicating. Problems arise when one person has trouble "translating" a message that comes from someone who uses a different style. There are at least four communication styles into which people tend to fit: the Intuitor, the Senser, the Thinker, and the Feeler. Of course, people may shift around or possess characteristics from more than one category, but for most people one or two styles are dominant. Recognizing specific styles in others will help you communicate more clearly.[6]

The Styles

The following are characteristics of each communication style.

A person using the *Intuitor* style is interested in ideas more than details, often moves from one concept or generalization to another without referring to examples, values insight and revelations, talks about having a vision, looks toward the future, and can be oriented toward the spiritual.

A person showing the style of *Senser* prefers details or concrete examples to ideas and generalizations, is often interested in the parts rather than the whole, prefers the here and now to the past or future, is suspicious of sudden insights or revelations, and feels that "seeing is believing."

A person using the *Thinker* style prefers to analyze situations, likes to solve problems logically, sees ideas and examples as useful if they help to figure something out, and becomes impatient with emotions or personal stories unless they have a practical purpose.

A person showing the style of *Feeler* is concerned with ideas and examples that relate to people, often reacts emotionally, is concerned with values and their effects on people and other living things, doesn't like "cold logic" or too much detail.

You can benefit from shifting from style to style according to the situation, particularly when trying to communicate with someone who prefers a style different from yours. Shifting, however, is not always easy or possible. The most important task is to try to understand the different styles and to help others understand yours. In general, no one style is any better than another. Each has its own positive effects that enhance communication and negative effects that can hinder it, depending on the situation.

Identifying Your Styles

These four styles are derived from the Myers-Briggs Type Indicator (MBTI). Because the learning style assessments are also in part derived from the MBTI, you will notice similarities between those assessments and these communication styles. Table 10-2 shows how some learning styles may correspond loosely to the communication styles. Not all individual learning styles within the assessments are mentioned, and the styles that are noted may correspond with different styles in different situations, but these matchups depict the most common associations. Finding where your learning styles fit may help you to determine your dominant communication style or styles.

Adjusting to the Listener's Style

When you are the speaker, you will benefit from an understanding of both your own style and the styles of your listeners. It doesn't matter how clear you think you are being if the person you are speaking to can't "translate" your

Table 10-2 Learning styles and communication styles.

COMMUNICATION STYLE	LEARNING STYLES INVENTORY	PATHWAYS TO LEARNING (MULTIPLE INTELLIGENCES)	PERSONALITY SPECTRUM
Intuitor	Theoretical, Holistic	Intrapersonal	Adventurer
Senser	Factual	Bodily-Kinesthetic	Organizer
Thinker	Linear	Logical-Mathematical	Thinker
Feeler	Reflective	Interpersonal	Giver

message by understanding your style. Try to take your listener's style into consideration when you communicate.

Following is an example of how adjusting to the listener can aid communication.

An intuitor-dominant instructor to a senser-dominant student: "Your writing isn't clear." The student's reply: "What do you mean?"

- *Without adjustment:* If the intuitor doesn't take note of the senser's need for detail and examples, he or she may continue with a string of big-picture ideas that might further confuse and turn off the senser. "You need to elaborate more. Try writing with your vision in mind. You're not considering your audience."
- *With adjustment:* If the intuitor shifts toward a focus on detail and away from his or her natural focus on ideas, the senser may begin to understand, and the lines of communication can open. "You introduced your central idea at the beginning but then didn't really support it until the fourth paragraph. You need to connect each paragraph's idea to the central idea. Also, not using a lot of examples for support makes it seem as though you are writing to a very experienced audience."

Adjusting to the Communicator's Style

As a facet of communication, listening is just as important as speaking. When you are the listener, try to stay aware of the communication style of the person who is speaking to you. Observe how that style satisfies, or doesn't satisfy, what a person of your particular style prefers to hear. Work to understand the speaker in the context of his or her style and translate the message into one that makes sense to you.

Following is an example of how adjusting to the communicator can boost understanding.

A feeler-dominant employee to a thinker-dominant supervisor: "I'm really upset about how you've talked down to me. I don't think you've been fair. I haven't been able to concentrate since our discussion and it's hurting my performance."

- *Without adjustment:* If the thinker becomes annoyed with the feeler's focus on emotions, he or she may ignore them, putting up an even stronger barrier between the two people. "There's no reason to be upset. I told you clearly and specifically what needs to be done. There's nothing else to discuss."
- *With adjustment:* If the thinker considers that emotions are dominant in the feeler's perspective, he or she could respond to those emotions in a way that still searches for the explanations and logic the thinker understands best: "Let's talk about how you feel. Please explain to me what has caused you to become upset, and we'll discuss how we can improve the situation."

Overcoming Communication Problems

Communication problems may occur when information is not clearly presented, or when those who receive information filter it through their own perspectives and interpret it in different ways. A few of the most common communication problems follow, along with strategies to help you combat them.

Problem: Unclear or Incomplete Explanation
Solution: Support Ideas with Examples

When you clarify a general idea with supporting examples that illustrate how it works and what effects it causes, you will help your receiver understand what you mean and therefore have a better chance to hold his or her attention.

For example, if you tell a friend to take a certain class, that friend might not take you seriously until you explain why. If you then communicate the positive effects of taking that class (progress toward a major, an excellent instructor, friendly study sessions), you may get your message across. The same principle applies to your attitude toward this course. If others communicate to you specific examples of how your work in the course will benefit your education, career, and personal life, you may be more likely to apply yourself.

Work situations benefit from explanation as well. As a supervisor, if you assign a task without explanation, you might get a delayed response or find mistakes in your employee's work. If, however, you explain the possible positive effects of the task, you'll have better results.

Problem: Attacking the Receiver
Solution: Send "I" Messages

When a conflict arises, often the first instinct is to pinpoint what someone else did wrong. "You didn't lock the door!" "You never called last night!" "You left me out!" Making an accusation, especially without proof, puts the other person on the defensive and shuts down the lines of communication.

Using "I" messages will help you communicate your own needs rather than focusing on what you think someone else did wrong or should do differently. "I felt uneasy when I came to work and the door was unlocked." "I became worried about you when I didn't hear from you last night." "I felt disappointed when I realized that I couldn't join the party." "I" statements soften the conflict by highlighting the *effects* that the other person's actions have had on you, rather than the person or the actions themselves. When you focus on your own response and needs, your receiver may feel more free to respond, perhaps offering help and even acknowledging mistakes.

If you often feel dissatisfied and tense after an exchange, you may benefit from focusing more on your own needs when you communicate. Translate your anger into an "I" statement before speaking. Ask the other person, "Can we decide together how to improve this situation? Here's how I feel about what has happened." Using "I" statements will bring better results.

Problem: Passive or Aggressive Communication Styles
Solution: Become Assertive

Among the three major communication styles—aggressive, passive, assertive—the one that conveys a message in the clearest, most productive way is the assertive style. The other two, while commonly used, throw the communication out of balance. An aggressive communicator often denies the receiver a chance to respond, while a passive communicator may have trouble getting the message out. Assertive behavior strikes a balance between aggression and passivity. If you can be an assertive communicator, you will be more likely to get your message across while assuring that others have a chance to speak as well. Table 10-3 compares some characteristics of each kind of communicator.

Aggressive communicators focus primarily on their own needs. They can become angry and impatient when those needs are not immediately satisfied. In order to become more assertive, aggressive communicators might try to take time to think before speaking, avoid ordering people around, use "I" statements, and focus on listening to what the other person has to say.

Passive communicators deny themselves the power that aggressive people grab. They focus almost exclusively on the needs of others instead of on their own needs, experiencing frustration and tension that remains unexpressed. In order to become more assertive, passive communicators might try to acknowledge anger or hurt more often, speak up when they feel strongly about something, realize that they have a right to make requests, and know that their ideas and feelings are as important as anyone else's.

Communication Success Strategies

The strategies on the following page can help improve your communication.

TERMS

Assertive Able to declare and affirm one's own opinions while respecting the rights of others to do the same.

Table 10-3 Aggressive, passive, and assertive styles.

AGGRESSIVE	PASSIVE	ASSERTIVE
Loud, heated arguing	Concealing one's own feelings	Expressing feelings without being nasty or overbearing
Physically violent encounters	Denying one's own anger	Acknowledging emotions but staying open to discussion
Blaming, name-calling, and verbal insults	Feeling that one has no right to express anger	Expressing self and giving others the chance to express themselves equally
Walking out of arguments before they are resolved	Avoiding arguments	Using "I" statements to defuse arguments
Being demanding: "Do this"	Being noncommittal: "You don't have to do this unless you really want to . . ."	Asking and giving reasons: "I would appreciate it if you would do this, and here's why . . ."

Think before you speak. Spoken too soon, ideas can come out sounding nothing like you intended them to. Taking time to think, or even rehearsing mentally, can help you choose the best combination of words. Think it through and get it right the first time.

Don't withhold your message for too long. One danger of holding back is that a problem or negative feeling may become worse. Speaking promptly has two benefits: (1) you solve the problem sooner, and (2) you are more likely to focus on the problem at hand than to spill over into other issues.

Communicate in a variety of ways, and be sensitive to cultural differences. Remember that words, gestures, and tones mean different things to different people.

Be clear, precise, and to the point. Say exactly what you need to say. Link your ideas to clear examples, avoiding any extra information that can distract.

Communication is extremely important for building and maintaining personal relationships. Explore the role those relationships play in who you are.

HOW DO YOUR PERSONAL RELATIONSHIPS DEFINE YOU?

The relationships you have with friends, family members, and significant others often take center stage. Jobs and schooling can come and go, but you rely on the people with whom you share your life.

In addition to being part of your life, the people around you help to define who you are. Since birth, you have learned by taking in information from verbal and nonverbal language. The chain of learning stretches back through time, each link formed by an exchange of information between people. Those with whom you live, play, study, and work are primary sources of ideas, beliefs, and ways of living. You grow and change as you have new experiences, evaluate them, and decide what to learn from them.

These influential relationships can affect other areas of your life. You have probably experienced conflict that caused you to be unable to sleep, eat, or get any work done. On the other hand, a successful relationship can have positive effects on your life, increasing your success at work or at school. Following are some strategies for improving your personal relationships.

Relationship Strategies

If you can feel good about your personal relationships, other areas of your life will benefit. Here are some suggestions.

Make personal relationships a high priority. Nurture the ones you have and be open to developing new ones. Life is meant to be shared. In some marriage ceremonies, the bride and groom share a cup of wine that symbolizes life. One of the reasons for this tradition is to double the sweetness of life by tasting it together, and another is to cut the bitterness in half by shar-

ing it. Any personal relationship can benefit from the experience of this kind of sharing.

Invest time. You devote time to education, work, and the other priorities in your life. Relationships need the same investment. They are like plants in a garden, needing nourishment to grow and thrive. Your attention provides that nourishment. In addition, spending time with people you like can relieve everyday stress and strain. When you make time for others, everyone benefits.

Spend time with people you respect and admire. Life is too short to hang out with people who bring you down, encourage you to participate in activities you don't approve of, or behave in ways that upset you. Develop relationships with people whom you respect, whose choices you admire, and who inspire you to be all that you can be. This doesn't mean that you have to agree with everything that others do. For example, you may disagree with a friend who lets his child watch a lot of TV. However, you may severely disapprove of someone who disciplines a child violently, and you may choose to end your association with that person.

Work through tensions. Negative feelings can multiply when left unspoken. Unexpressed feelings about other issues may cause you to become disproportionately angry over a small issue. A small annoyance over dishes in the sink can turn into a gigantic fight about everything under the sun. Get to the root of the problem. Discuss it, deal with it, and move on.

Refuse to tolerate violence. It isn't easy to face the problem of violence or to leave a violent relationship. People may tolerate violence out of a belief that it will end, a desire to keep their families together, a self-esteem so low that they believe they deserve what they get, or a fear that trying to leave may lead to greater violence. No level of violence is acceptable. Someone who behaves violently toward you cannot possibly have your best interests at heart. If you find that you are either an aggressor or a victim, do your best to get help.

Show appreciation. In this fast-moving world, people don't thank each other often enough. If you think of something positive, say it. Thank someone for a service, express your affection with a smile. A little positive reinforcement goes a long way toward nurturing a relationship.

If you want a friend, be a friend. The Golden Rule, "Do unto others as you would have them do unto you," never goes out of style. If you treat a friend with the kind of loyalty and support that you appreciate yourself, you are more likely to receive the same in return.

Take risks. It can be frightening to reveal your deepest dreams and frustrations, to devote yourself to a friend, or to fall in love. You can choose not to reveal yourself or give yourself to a friendship at all. However, giving is what feeds a relationship, bringing satisfaction and growth. If you take the plunge, you risk disappointment and heartbreak, but you stand to gain the incredible benefits of companionship, which for most people outweigh the risks.

Keep personal problems in their place. Solve personal problems with the people directly involved and no one else. If at all possible, try not to bring your emotions into class or work. Doing so may hurt your performance while doing nothing to help your problem. If you are overwhelmed by a personal problem, try to address it before you go to class or work. If it's impossible to address it at that time, at least make a plan that you can carry out later. Making some step toward resolving the problem will help you concentrate on other things.

If it doesn't work out, find ways to cope. Everyone experiences strain and breakups in intimate relationships, friendships, and family ties. Be kind to yourself and use coping strategies that help you move on. Some people need lots of time alone; others need to spend time with their friends and family. Some seek more formal counseling. Some people throw their energy into a project, a job, a class, a new workout regimen, or anything else that will take their mind off what hurts. Some just need to cry it out and be miserable for a while. Some write in a journal or write letters to the person that they never mail. Do what's right for you, and believe that sooner or later you can emerge from the experience stronger and with new perspective.

Now and again, your personal relationships will experience conflict. Following are ideas for how to deal with conflict and criticism in a productive and positive way.

HOW CAN YOU HANDLE CONFLICT AND CRITICISM?

Conflict and criticism, as unpleasant as they can often be, are natural elements in the dynamic of getting along with others. It's normal to want to avoid people or situations that cause distress. However, if you can face your fears and think through them critically, you can gain valuable insight into human nature—your own and that of others. You may be able to make important changes in your life based on what you learn.

Conflict Strategies

Conflicts both large and small arise when there is a clash of ideas or interests. You may have small conflicts with a housemate over food left out overnight, a door left unlocked, or a bill that needs paying. On the other end of the spectrum, you might encounter major conflicts with your partner about finances, with an instructor about a failing grade, or with a person who treats you unfairly because of your race, gender, age, or ethnic origin.

Conflict can create anger and frustration, shutting down communication. The two most destructive tendencies are to avoid the conflict altogether (a passive tactic) or to let it escalate into a blowout fight (an aggressive tendency). Avoidance doesn't make the problem go away—in fact, it will probably worsen. If you tend to be passive, assert yourself by acknowledging and expressing your feelings as soon as you can put them into words. On the other hand, a shouting match gives no one an opportunity or desire to listen. If you tend to be aggressive, give yourself time to cool down before you address a conflict. Try to express what you feel without letting your emotions explode.

If calmly and intelligently handled, conflict can shed light on new ideas and help to strengthen bonds between those involved. The primary keys to conflict resolution are calm communication and critical-thinking skills. Think through any conflict using what you know about problem solving.

Identify and analyze the problem. Determine the severity of the problem by looking at its effects on everyone involved. Then, find and analyze the causes of the problem.

Brainstorm possible solutions. Consider as many angles as you can, without judgment. Explore what ideas you can come up with from what you or others have done in a similar situation.

Explore each solution. Evaluate the positive and negative effects of each solution. Why might each work, or not work, or work partially? What would take into account everyone's needs? What would cause the least stress? Make sure everyone has a chance to express an opinion.

Choose, carry out, and evaluate the solution you decide is best. When you have implemented your choice, evaluate its effects. Decide whether you feel it was a good choice.

One more hint: Use "I" statements. Focus on the effect the problem has had on you rather than focusing on someone who caused it. Show that you are taking responsibility for your role in the exchange.

Dealing With Criticism and Feedback

Feedback
Evaluative or corrective information about an action or process.

No one gets everything right all the time. People use constructive criticism and feedback to communicate what went wrong and to suggest improvements. Consider any criticism carefully. If you always interpret criticism as a threat, you will close yourself off from learning. Even if you eventually decide that you disagree, you can still learn from exploring the possibility. Know that you are strong enough to embrace criticism and become a better person because of it.

Criticism can be either constructive or unconstructive. Criticism is considered constructive when it is offered supportively and contains useful suggestions for improvement. On the other hand, *unconstructive* criticism focuses on what went wrong, doesn't offer alternatives or help, and is often delivered in a negative or harsh manner. Whereas constructive criticism can promote a sense of hope for improvement in the future, unconstructive criticism can create tension, bad feelings, and defensiveness.

Constructive
Promoting improvement or development.

Any criticism can be offered constructively or unconstructively. Consider a case where someone has continually been late to work. A supervisor can offer criticism in either of these ways:

Constructive: The supervisor talks privately with the employee. "I've noticed that you have been late to work a lot. Other people have had to do some of your work. Is there a problem that is keeping you from being on time? Is it something that I or someone else can help you with?"

Unconstructive: The supervisor watches the employee slip into work late. The supervisor says, in front of other employees, "Nice to see you could make it. If you can't start getting here on time, I might look for someone else who can."

If you can learn to give constructive criticism and deal with whatever criticism comes your way from others, you will improve your relationships and your productivity. When offered constructively and carefully considered, criticism can bring about important changes.

Giving Constructive Criticism

When you offer criticism, use the following steps to communicate clearly and effectively:

1. *Criticize the behavior rather than the person.* In addition, make sure the behavior you intend to criticize is changeable. Chronic lateness can be changed; a physical inability to perform a task cannot.
2. *Define specifically the behavior you want to change.* Try not to drag any side issues into the conversation.
3. *Balance criticism with positive words.* Alternate critical comments with praise in other areas.
4. *Stay calm and be brief.* Avoid threats, ultimatums, or accusations. Use "I" messages; choose positive, nonthreatening words, so the person knows that your intentions are positive.
5. *Explain the effects caused by the behavior that warrants the criticism.* Help the person understand why a change needs to happen, and talk about options in detail. Compare and contrast the effects of the current behavior with the effects of a potential change.
6. *Offer help in changing the behavior.* Lead by example.

> "Do not use a hatchet to remove a fly from your friend's forehead."
> CHINESE PROVERB

Receiving Criticism

When you find yourself on the receiving end of criticism, use these coping techniques:

1. *Listen to the criticism before you speak up.* Resist the desire to defend yourself until you've heard all the details. Decide if the criticism is offered in a constructive or unconstructive manner.
2. *Think the criticism through critically.* Evaluate it carefully. While some criticism may come from a desire to help, other comments may have less honorable origins. People often criticize others out of jealousy, anger, frustration, or displaced feelings. In cases like those, it is best (though not always easy) to let the criticism wash right over you.
3. *If it is unconstructive, you may not want to respond at that moment.* Unconstructive criticism can inspire anger that might be destructive to

express. Wait until you cool down and think about the criticism to see if there is anything important hiding under how it was presented. Then, tell the person that you see the value of the criticism, but also communicate to him or her how the delivery of the criticism made you feel. If he or she is willing to talk in a more constructive manner, continue with the following steps below. If not, your best bet may be to consider the case closed and move on.

4. *If it is constructive, ask for suggestions of how to change the criticized behavior.* You could ask, "How would you handle this if you were in my place?"
5. *Before the conversation ends, summarize the criticism and your response to it.* Repeat it back to the person who offered it. Make sure both of you understand the situation in the same way.
6. *If you feel that the criticism is valid, plan a specific strategy for correcting the behavior.* Think over what you might learn from changing your behavior. If you don't agree with the criticism even after the whole conversation, explain your behavior from your point of view.

Remember that the most important feedback you will receive in school is from your instructors, and the most important on-the-job feedback will come from your supervisors, more experienced peers, and occasionally clients. Making a special effort to take in this feedback and consider it carefully will help you learn many important lessons. Even when the criticism is not warranted, the way you respond is important. Furthermore, knowing how to handle conflict and criticism will help you define your role and communicate with others when you work in groups.

WHAT ROLE DO YOU PLAY IN GROUPS?

Group interaction is an important part of your educational, personal, and working life. With a team project at work or a cooperative learning exercise in school, for example, being able to work well together is necessary in order to accomplish a goal.

The two major roles in the group experience are those of *participant* and *leader*. Any group needs both in order to function successfully. Become aware of the role you tend to play when relating to others. Try different roles to help you decide where you can be most effective. The following strategies (from *Contemporary Business Communication*, by Louis E. Boone, David L. Kurtz, and Judy R. Block) are linked to either participating or leading.[6]

Being an Effective Participant

Some people are happiest when participating in group activities that someone else leads and designs. They don't feel comfortable in a position of control or having the power to set the tone for the group as a whole. They trust others to make those decisions, preferring to help things run smoothly by taking on an assigned role in the project and seeing it through. Participators need to remember that they are "part owners" of the process. Each team member has a responsibility for, and a stake in, the outcome. The following strategies will help a participant to be effective.

Participation Strategies

Get involved. If a decision you don't like is made by a group of which you are a member and you stayed uninvolved in the decision, you have no one to blame but yourself for not speaking up. Put some energy into your participation and let people know your views. You are as important a team member as anyone else, and your views are likewise valuable.

Be organized. When you participate with the group as a whole, or with any of the team members, stay focused and organized. The more organized your ideas are, the more people will listen, take them into consideration, and be willing to try them.

Be willing to discuss. Everyone has an equal right to express his or her ideas. Even as you enthusiastically present your opinions, be willing to consider those of others. Keep an open mind and think critically about other ideas before you assume they won't work. If a discussion heats up, take a break or let a more neutral group member mediate.

Keep your word. Make a difference by doing what you say you're going to do. Let people know what you have accomplished. If you bring little or nothing to the process, your team may feel as if you weigh them down.

Focus on ideas, not people. One of the easiest ways to start an argument is for participants to attack group members themselves instead of discussing their ideas. Separate the person from the idea, and keep the idea in focus.

Play fairly. Give everyone a chance to participate. Be respectful of other people's ideas. Don't dominate the discussion or try to control or manipulate others.

Being an Effective Leader

Some people prefer to initiate the action, make decisions, and control how things proceed. They have ideas they want to put into practice and enjoy explaining them to others. They are comfortable giving direction to people and guiding group outcomes. Leaders often have a big-picture perspective; it allows them to see how all of the different aspects of a group project can come together. In any group setting, the following strategies will help a leader succeed.

Leadership Strategies

Define and limit projects. One of the biggest ways to waste time and energy is to assume that a group will know its purpose and will limit tasks on its own. A group needs a leader who can define the purpose of the gathering and limit tasks so the group doesn't take on too much. Some common purposes are giving/exchanging information, brainstorming, making a decision, delegating tasks, or collaborating on a project.

Map out who will perform which tasks. A group functions best when everyone has a particular contribution to make. You don't often choose who you work with—in school, at work, or in your family—but you can help different personalities work together by exploring who can do what best. Give people specific responsibilities, and trust that they will do their jobs.

Set the agenda. The leader is responsible for establishing and communicating the goal of the project and how it will proceed. Without a plan, it's easy to get off track. Having a written agenda to which group members can refer is helpful. A good leader invites advice from others when determining group direction.

Focus progress. Even when everyone knows the plan, it's still natural to wander off the topic. The leader should try to rein in the discussion when necessary, doing his or her best to keep everyone to the topic at hand. When challenges arise midstream, the leader may need to help the team change direction.

Set the tone. Different group members bring different attitudes and mental states to a gathering. Setting a positive tone helps to bring the group together and motivate people to peak performance. When a leader values diversity in ideas and backgrounds and sets a tone of fairness, respect, and encouragement, group members may feel more comfortable contributing their ideas.

Evaluate results. The leader should determine whether the team is accomplishing its goals. If the team is not moving ahead, the leader needs to make changes and decisions.

If you don't believe you fit into the traditional definition of a leader, remember that there are other ways to lead that don't involve taking charge of a group. You can lead others by setting an honorable example in your actions, choices, or words. You can lead by putting forth an idea that takes a group in a new direction. You can lead by being the kind of person whom others would like to be.

It takes the equal participation of all group members to achieve a goal. Whatever role works best for you, know that your contribution is essential. You may even play different roles with different groups, such as if you were a participator at school and a leader in a self-help group. Finally, stay aware of group dynamics; they can shift quickly and move you into a new position you may or may not like. If you don't feel comfortable, speak up. The happier each group member is, the more effectively the group as a whole will function.

Kente

The African word *kente* means "that which will not tear under any condition." *Kente* cloth is worn by men and women in African countries such as Ghana, Ivory Coast, and Togo. There are many brightly colored patterns of *kente*, each beautiful, unique, and special.

Think of how this concept applies to being human. Like the cloth, all people are unique, with brilliant and subdued aspects. Despite any mistreatment or misunderstanding by the people you encounter in your life, you need to work to remain strong so that you don't tear and give way to disrespectful behavior. This strength can help you to endure, stand up against any injustice, and fight peacefully but relentlessly for the rights of all people.

Chapter 10 Applications

Name _____ Date _____

KEY INTO YOUR LIFE
Opportunities to Apply What You Learn

 Diversity Discovery

Express your own personal diversity. Describe yourself in response to the following questions.

What ethnic background(s) do you have? _Hispanic_

Name one or more facts about you that someone wouldn't know from simply looking at you. _that I have 3 tattoos_

Name two values or beliefs that govern how you live, what you pursue, and/or with whom you associate. _family and education — I believe make you a better person._

What other characteristics or choices define your uniqueness? _I am respectful._

Now, join with a partner in your class. Try to choose someone you don't know well. Your goal is to communicate what you have written to your partner, and for your partner to communicate to you in the same way. Spend ten minutes talking together, and take notes on what the other person says. At the end of that period, join together as a class. Each person will describe his or her partner to the class.

What did you learn about your partner that surprised you? _____

What did you learn that went against any assumptions you may have made about that person based on his or her appearance, background, or behavior?

Has this exercise changed the way you see this person or other people? Why or why not? _____

10.2 Your Communication Style

Look back at the four styles on pp. 248–249: Intuitor, Thinker, Feeler, and Senser. Which describes you the best? Rank the four styles, listing first the one that fits most, and listing last the one that fits least.

1. thinker
2. feeler
3. senser
4. Intuitor

Of the two styles that best fit you, which one has more positive effects on your ability to communicate? What are those effects? feeler. personal stories may sometimes help you relate better with others.

Which style has more negative effects? What are they? intuitor. I can not be a spiritual person. Anyone start to talk to my about religion, I just don't know what to say

To determine whether you are primarily passive, aggressive, or assertive, read the following sentences and circle the ones that sound like something you would say to a peer.

1. Get me the keys.
2. *Would you mind if I stepped out just for a second?* ⭕
3. *Don't slam the door.* ⭕
4. I'd appreciate it if you would have this done by two o'clock.
5. I think maybe it needs a little work just at the end, but I'm not sure.
6. *Please take this back to the library.* ⭕
7. You will have a good time if you join us.
8. Your loss.
9. I don't know, if you think so. I'll try it.
10. *Let me know what you want me to do.* ⭕
11. Turn it this way and see what happens.
12. We'll try both our ideas and see what works best.
13. I want it on my desk by the end of the day.
14. Just do what I told you.
15. If this isn't how you wanted it to look, I can change it. Just tell me and I'll do it.

Aggressive communicators would be likely to use sentences 1, 3, 8, 13, and 14.
Passive communicators would probably opt for sentences 2, 5, 9, 10, and 15.
Assertive communicators would probably choose sentences 4, 6, 7, 11, and 12.

In which category did you choose the most sentences? __Passive__

If you scored as an assertive communicator, you are on the right track. If you scored in the aggressive or passive categories, analyze your style. What are the effects? Give an example in your own life of the effects of your style.

I may experience frustration and not say anything.

Turn back to p. 252 to review suggestions for aggressive or passive communicators. What can you do to improve your skills?

Speak up. Grab some power.

Problem Solving Close to Home

Divide into small groups of two to five students. Assign one group member to take notes. Discuss the following questions, one by one:

1. What are the three largest problems my school faces with regard to how people get along with and accept others?
2. What could my school do to deal with the three problems listed above?
3. What can each individual student do to deal with the three problems listed above? (Talk about what you specifically feel that you can do.)

When you are finished, gather as a class. Each group should share their responses with the class. Observe the variety of problems and solutions. Notice whether more than one group came up with one or more of the same problems. You may want to assign one person in the class to gather all of the responses together. That person, together with your instructor, could put these responses into an organized document that you can share with the upper-level administrators at your school.

KEY TO SELF-EXPRESSION
Discovery Through Journal Writing

To record your thoughts, use a separate journal or the lined pages at the end of the chapter.

New Perspective[7]

Imagine that you have no choice but to change either your gender (male/female) or your racial/ethnic/religious group. Which would you change, and why? What do you anticipate would be the positive and negative effects of the change—in your social life, in your family life, on the job, at school? How would what you know and experience before the change affect how you would behave after the change?

Name _____ Date _____ **Journal**

Journal

Name _____ Date _____

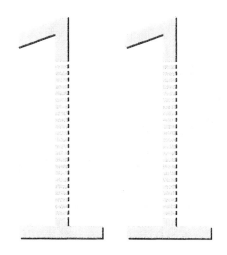

Managing Career and Money

Reality Resources

Many people either love their jobs but don't make much money, or dislike their jobs but are paid well. Still other people have neither job satisfaction nor a good paycheck to show for their work. The most ideal career interests and challenges you *and* pays you enough to live comfortably.

Career exploration, job-hunting strategy, and money management can work together to help you find that ideal career. In this chapter, you will first look at career exploration and how to balance work and school. Then you will explore how to bring in money with financial aid and how to manage the money you have. Managing your resources and investigating career options can help you develop skills and insights that will serve you throughout your life.

In this chapter, you will explore answers to the following questions:

- How can you plan your career?
- How can you juggle work and school?
- What should you know about financial aid?
- How can strategic planning help you manage money?
- How can you create a budget that works?

How Can You Plan Your Career?

College is an ideal time to investigate careers, because so many different resources are available to you. Students are in all different stages of thought when it comes to careers. You may not have thought too much about it yet. You may have already had a career for years and are looking for a change. You may have been set on a particular career but are now having second thoughts. Regardless of your starting point, now is the time to make progress.

Even outside this particular chapter, everything you read and work on in this book is geared toward college and workplace success. As you work on the exercises, you think critically, become a team player, hone writing skills, and develop long-term planning, all of which prepare you to thrive in any career as well as in your studies.

Define a Career Path

Aiming for a job in a particular career area requires planning the steps that can get you there. Whether these steps take months or years, they help you focus your energies on your goal. Defining a career path involves investigating yourself, exploring potential careers, and building knowledge and experience.

Investigate

When you explore your learning style in Chapter 3, evaluate your ideal note-taking system in Chapter 7, or look at how you relate to others in Chapter 10, you build self-knowledge. Gather everything that you know about yourself, from this class or from any of your other life experiences, and investigate. What do you know or do best? Out of jobs you've had, what did you like and not like to do? How would you describe your personality? And finally, what kinds of careers make the best use of everything you are?

Don't feel as though you should automatically know what you want to do. Most students who have not been in the workplace don't know what career they want to pursue. Students who have been working often return to school to explore other careers that they might prefer. More and more, people are changing careers many times in their lives instead of sticking with one choice. This discovery is a lifelong process.

The potential for change applies to majors as well. If you declare a major and decide later that you don't like it, feel glad that you were able to discover that fact about yourself.

Explore Potential Careers

Your school's career center is an important resource in your investigation of career opportunities. The career center may offer job listings, occupation lists, assessments of skills and personality types, questionnaires to help you pinpoint career areas that may suit you, informational material about different career areas, and material about various companies. The people who work at the center can help you sort through the material.

Look at Table 11-1 for some of the kinds of questions you might ask as you talk to people, such as instructors, relatives, and fellow students, about careers or investigate materials. Use your critical-thinking skills to broaden your question-asking beyond just what tasks you perform for any given job. Many other factors will be important to you.

Within every career field, a wide array of job possibilities exists that you might not see right away. For example, the medical world involves more than just doctors and nurses. Emergency medical technicians respond to emergencies, administrators run hospitals, researchers test new drugs, lab technicians administer specific procedures such as X-rays, pharmacists administer prescriptions, retirement community employees work with the elderly, and more.

Within each job, there is also a variety of tasks and skills that often go beyond what you know. You may know that an instructor teaches, but you may not see that instructors also often write, research, study, create course outlines, create strategy with other instructors, give presentations, and counsel. Push past your first impression of any career and explore what else it entails. Expand your choices as much as you can using thorough investigation and an open mind.

Table 11-1 Critical thinking questions for career investigation.

What can I do in this area that I like/am good at?	Do I respect the company and/or the industry?
What are the educational requirements (certificates or degrees, courses)?	Do companies in this industry generally accommodate special needs (child care, sick days, flex time, or working at home)?
What skills are necessary?	Can I belong to a union?
What wage or salary is normal for an entry-level position, and what benefits can I expect?	Are there opportunities in this industry within a reasonable distance from where I live?
What kinds of personalities are best suited to this kind of work?	What other expectations are there beyond the regular workday (travel, overtime, etc.)?
What are the prospects for moving up to higher-level positions?	Do I prefer a service or manufacturing industry?

Build Knowledge and Experience

Having knowledge and experience specific to the career area you want to pursue will be valuable on the job hunt. Courses, internships, jobs, and volunteering are four great ways to build both (see also Chapter 2).

Courses. When you narrow your career exploration to a couple of areas that interest you, look through your school course catalog and take a course or two in those fields. How you react to these courses will give you important clues as to how you feel about the area in general. Be careful to evaluate your experience based on how you feel about the subject matter and not other factors. Think critically. If you didn't like a course, what was the cause: an instructor you didn't like, a time of day when you tend to lose energy, or truly a lack of interest in the material?

Internships. An internship may or may not offer pay. While this may be a financial drawback, the experience you can gather and contacts you can make may be worth the work. Many internships take place during the summer, but some part-time internships are also available during the school year. Companies that offer internships are looking for people who will work hard in exchange for experience you can't get in the classroom.

Absorb all the knowledge you can while working as an intern. If you discover a career worth pursuing, you'll have the internship experience behind you when you go job hunting. Internships are one of the best ways to show a prospective employer some "real world" experience and initiative.

Jobs. No matter what you do for money while you are in college, whether it is in your area of interest or not, you may discover career opportunities that appeal to you. Someone who takes a night-shift nursing assistant job to make extra cash might discover an interest in health sciences research. Someone who answers phones for a newspaper company might be drawn into science writing. Stay aware of the possibilities around you.

Volunteering. Offering your services in the community or at your school can introduce you to career areas and increase your experience. Some schools have programs that can help you find opportunities to work as an aid on campus or volunteer off campus. Recently, certain schools have even begun listing volunteer activities on student transcripts. Find out what services your school offers. Volunteer activities are important to note on your resume. Many employers seek candidates who have shown commitment through volunteering.

Map Out Your Strategy

After you've gathered enough information to narrow your career goals, plan strategically to achieve them. Make a career time line that illustrates the steps toward your goal, as shown in Figure 11-1. Mark years and half-year points (and months for the first year), and write in the steps where you think they should take place. If your plan is five years long, indicate what you plan to do by the fourth, third, and second years, and then the first year, including a six-month goal and a one-month goal for that first year. Set goals that establish

---

TERMS

Internship
A temporary work program in which a student can gain supervised practical experience in a particular professional field.

1 month	Enter community college on part-time schedule	**Figure 11-1**
3 months		
6 months	Meet with advisor to discuss desired major and required courses	Career time line.
1 year		
	Declare major in computer science	
2 years	Switch to full-time class schedule	
3 years	Graduate with associate's degree	
	Transfer to 4-year college	
4 years	Work part-time in college's computer lab	
5 years	Student internship at Microsoft	
	Graduate with bachelor's degree in computer science	
6 years	Get a job at Microsoft and then apply to graduate school	

who you will talk to, what courses you will take, what skills you will work on, what jobs or internships you will investigate, and any other research you need to do. Your path may change, of course—use your time line as a guide rather than as a rigid plan.

Know What Employers Want

Certain basic skills will make you an excellent job candidate no matter what career you decide to pursue. Employers look for particular skills and qualities that signify an efficient and effective employee. You can continue to develop these skills as you work in current and future jobs—and you will, if you always strive to improve.

Communication skills. Being able to listen well and express yourself in writing and speaking is a key to workplace success. Much can be accomplished through efficient, open communication. Being able to adjust to different communication styles is an important factor.

Problem solving. Any job will present problems that need to be solved. An employee who knows how to assess any situation and apply the problem-solving process to it will stand out.

Decision making. Decisions large and small are made in every workplace every day. Knowing how to think through and make decisions will help you in any job.

"Whatever you think you can do or believe you can do, believe it. Action has magic, grace, and power in it."

JOHANN WOLFGANG VON GOETHE

Teamwork. It is a rare workplace that has only one employee, and even then, that person will interact with different kinds of people on the phone or through a computer. The importance of being able to work well with others cannot be overemphasized. If there is a weak link in any team, the whole company suffers.

Multicultural communication. The workplace is becoming increasingly diverse. The more you can work well with people different from yourself and open your mind to their points of view, the more valuable an employee you will be.

Leadership. The ability to influence others in a positive way will earn you respect and keep you in line for promotions. Taking the lead will often command attention.

Creativity. When you can see the big picture as well as the details and can let your mind come up with unexpected new concepts and plans, you will bring valuable suggestions to your workplace.

Commitment. You will encounter many difficult situations at work. The ability to continue to work hard through such situations is extremely important. In addition, if you introduce a new and creative idea, you can gain support for it through having a strong commitment to it yourself.

Values and integrity. Your personal values and integrity will help guide everything you do. In your actions and decisions, consider what you value and what you believe is right.

These skills appear throughout this book, and they are as much a part of your school success as they are of your work success. The more you develop them now, the more employable and promotable you will prove yourself to be. You may already use them on the job if you are a student who works, which we will discuss next.

HOW CAN YOU JUGGLE WORK AND SCHOOL?

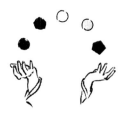

What you are studying today can prepare you to find a job, when you graduate, that suits your abilities and brings in enough money to support your needs and lifestyle choices. In the meantime, though, you can make work a part of your student life in order to make money, explore a career, and/or increase your future employability through contacts or resume building.

As the cost of education continues to rise, more and more students are working and taking classes at the same time. In the school year 1995–96, 79 percent of undergraduates worked while in school. Most student workers 23 years of age or younger worked part-time jobs (37.8 percent). Of students over the age of 23, a majority worked full-time jobs (nearly 55 percent).[1]

Being an employed student isn't for everyone. Adding a job to the list of demands on your time and energy may create problems if it sharply reduces study time or family time. However, many people want to work and many need to work in order to pay for school. Ask yourself important questions about why or why not to work. Weigh the potential positive and negative effects of working. From those answers you can make a choice that you feel benefits you most.

Effects of Working While in School

Working while in school has many different positive and negative effects, depending on the situation. Evaluate any job opportunity by looking at these effects. Following are some that might come into play.

Potential Positive Effects

Money earned. A job can provide crucial income to pay for rent, transportation, food, and important bills. It may even help you put some savings away in order to create a financial cushion.

General and career-specific experience. Important learning comes from hands-on work. Your education "in the trenches" can complement your classroom experience. Even if you don't work in your chosen field, you can improve universal skills such as teamwork and communication.

Being able to keep a job you currently hold. If you leave a job temporarily, your company might not be able to hold your position open until you come back. Consider adjusting your responsibilities or hours while still holding down your job.

Keeping busy. Work can provide a stimulating break from studying. In fact, working up to fifteen hours a week may actually enhance academic performance, because working students often manage their time more efficiently and may gain confidence from their successes in the workplace. Working on campus may help you manage your time and connect to your school experience.

Potential Negative Effects

Time commitment. A nonworking student splits their time between academic and personal life, but a working student must add a third, time-consuming factor. More responsibilities and less time to fulfill them demand more efficient time management. Many schools recommend that students work a maximum of twenty hours a week while taking a full course load.

Adjusting priorities. The priority level of your job may vary. For a student who depends on the income, work may take priority over study time. Evaluate priorities carefully. Realize that you may have to reduce social activities, exercise at home, cut back on nonacademic activities, or lighten your course load in order to maintain a job and still get studying done. Your job is important, but if you are also committed to school, earning a good GPA may be just as crucial.

Shifting gears. Unless your job meshes with your classroom curriculum, it may take some effort to shift gears mentally as you move back and forth between academia and the workplace. Each environment has its own set of people, responsibilities, joys, and problems. Establish mental boundaries that can help you shake off academic stress while at work, and vice versa.

If you consider the positive and negative effects and decide that working will benefit you, you should establish what you need in a job (see Table 11-2).

Sources of Job Information

Many different routes can lead to satisfying jobs. Use your school's career planning and placement office, networking skills, classified ads, employment agencies, and on-line services to help you explore.

Your School's Career Planning and Placement Offices

Generally, the career planning and placement office deals with post-graduation job placements, while the student employment office, along with the financial aid office, has more information about working while in school. At either location you might find general workplace information, listings of job opportunities, sign-ups for interviews, and contact information for companies. The career office may hold frequent informational sessions on different topics. Your school may also sponsor job or career fairs that give you a chance to explore job opportunities.

Many students, because they don't seek job information until they're about to graduate, miss out on much of what the career office can do. Don't wait until the last minute. Start exploring your school's career office early in your university life. The people and resources there can help you at every stage of your career and job exploration process.

Networking

Networking is one of the most important job-hunting strategies. With each person you get to know, you build your network and tap into someone else's. Imagine a giant think link connecting you to a web of people just a couple of phone calls away. Of course, not everyone with whom you network will come through for you. Keep in contact with as many people as possible in the hope that someone will. You never know who that person might be.

With whom can you network? Friends and family members may know of jobs or other people who can help you. At your school, instructors, administrators, or counselors may give you job or contact information. Attend professional conferences by contacting the on-campus club or state chapter (see appendix). Some schools even have opportunities for students to interact with alumni. Look to your present and past work experience for more leads. Employers or co-workers may know someone who needs new employees. A former employer might even hire you back with similar or adjusted hours, if you left on good terms.

The contacts with whom you network aren't just sources of job opportunities. They are people with whom you can develop lasting, valuable relationships. They may be willing to talk to you about how to get established,

TERMS

Networking
The exchange of information and/or services among individuals, groups, or institutions.

TERMS

Contact
A person who serves as a carrier or source of information.

Table 11-2 What you may need in a job.

NEED	DESCRIPTION
Salary/wage level	Consider how much money you need to make month by month and yearly. You may need to make a certain amount for the year as a whole, but you may need to earn more of that total amount during the months when you are paying tuition. Consider also the amount that justifies taking the time to work. If a job pays well but takes extra hours that should go toward studying or classes, it might not be worth it. Take time to compare the positive effects with the negative effects of any job's pay structure.
Time of day	When you can work depends on your school schedule. For example, if you take classes Monday, Tuesday, and Thursday during the day, you could look for a job with weekend or evening hours. If you attend evening classes, a daytime job could work fine.
Hours per week (part-time vs. full-time)	If you take classes part-time, you may choose to work a full-time job. If you are a full-time student, it may be best to work part-time. Balance your priorities so that you can accomplish your schoolwork and still make the money you need.
Duties performed	If you want hands-on experience in your chosen field, narrow your search to jobs that can provide it. On the other hand, if a regular paycheck is your priority, you might not care as much about what you do. Consider if there is anything you absolutely hate to do. Working somewhere and/or doing something that makes you miserable may not be worth any amount of money.
Location	Weigh the effects of how long it takes to get to a job against what you are getting out of it, and decide whether it is worth your while. A job at or near your school may give unparalleled convenience. When you know you can get to work quickly, you can schedule your day more tightly and get more done.
Flexibility	Even if your classes are at regular times, you might have other projects and meetings at various times. Do you need a job that offers flexibility, allowing you to shift your working time when you have to attend to an academic or family responsibility that takes priority? Choose according to the flexibility you require.
Affiliation with school or financial aid program	Some financial aid packages, especially if they involve funds from your school, can require you to take work at the school or a federal organization. In that case you would have to choose among the opportunities offered.
Accommodation of special needs (Americans with Disabilities Act)	If you have a hearing or vision impairment, reduced mobility, or other special needs, employers must accommodate them.

the challenges on the job, what they do each day, how much you can expect to make, or any other questions you have similar to those in Table 11-1. Thank your contacts for their help and don't forget them. Networking is a two-way street. Even as you receive help, be ready to extend yourself to others who may need help and advice from you.

Classified Ads

Some of the best job listings are in daily or periodic newspapers. Most papers print help wanted sections in each issue, organized according to career field categories. At the beginning of most help wanted sections you will find an index that tells you the categories and on what pages they begin in the listings. Individual ads describe the kind of position available and will give a telephone number or post office box for you to contact. Some ads will include additional information such as job requirements, a contact person, and the salary or wages offered.

You can run your own classified ads if you have a skill you want to advertise. Many college students make extra cash by doing specific tasks for campus employees or other students, such as typing, editing, cleaning, tutoring, or baby-sitting. You may want to advertise your particular job skills in your school or local paper.

On-Line Services

The Internet is growing as a source of job listings. Through it you can access job search databases such as the Career Placement Registry and U. S. Employment Opportunities. Web sites such as CareerPath.com and CareerMosaic list all kinds of positions. Individual associations and companies may also post job listings and descriptions, often as part of their World Wide Web pages. For example, IBM includes job openings on its Web page.

Employment Agencies

Employment agencies are organizations that help people find work. Most employment agencies will put you through a screening process that consists of an interview and one or more tests in your area of expertise. For example, someone looking for secretarial work may take a word-processing test and a spelling test, while someone looking for accounting work may take accounting and math tests. If you pass the tests and interview well, they will try to place you in a job.

Most employment agencies specialize in particular career or skill areas, such as accounting, medicine, legal, computer operation, graphic arts, child care, and food services. Agencies may place job seekers in either part-time or full-time employment. Many agencies also place people in temporary jobs, which can work well for students who are available from time to time. Such agencies may have you call in whenever you are free and will see if anything is available that day or week.

Employment agencies are a great way to hook into job networks. However, they usually require a fee that either you or the employer has to pay. Investigate any agency before signing on. See if your school's career counselors know anything about the agency, or if any fellow students have used it successfully. Ask questions so that you know as much as possible about how the agency operates.

Making a Strategic Job Search Plan

When you have gathered information on the jobs you want, formulate a plan for pursuing them. Organize your approach according to what you need to do and how much time you have to devote to your search. Do you plan to make three phone calls per day? Will you fill out three job applications a week for a month? Keep a record—on 3-by-5-inch cards, a computer file, or in a notebook—of the following:

- People you contact
- Companies to which you apply
- Jobs you rule out (for example, jobs that become unavailable or which you find out don't suit your needs)
- Response from your communications (phone calls to you, interviews, written communications) and the information on whomever contacted you (names, titles, times, and dates)

Keeping accurate records will enable you to both chart your progress and maintain a clear picture of the process. You never know when information might come in handy again. If you don't get a job now, another one could open up at the same company in a couple of months. In that case, well-kept records would enable you to contact key personnel quickly and efficiently. See Figure 11-2 for a sample file card.

Your Resume and Interview

Information on resumes and interviews fills many books. Therefore, your best bet is to consult some that will go into more detail, such as *The Resume Kit*, by Richard Beatty, or *Job Interviews for Dummies* by Joyce Lain Kennedy (don't be insulted by the title; it has lots of terrific information).

Figure 11-2

Sample file card.

Job/company: Child-care worker at Morningside Day Care
Contact: Sally Wheeler, Morningside Day Care, 17 Parkside Road, Silver Spring, MD 20910
Phone/fax/e-mail: (301) 555-3353 phone, (301) 555-3354 fax, no e-mail
Communication: Saw ad in paper, sent résumé and cover letter on October 7
Response: Call from Sally to set up interview

Interview on Oct. 15 at 2 p.m., seemed to get a positive response, she said she would contact me again by the end of the week

Follow-up: Sent thank-you note on October 16

The following basic tips can get you started on giving yourself the best possible chance at a job.

Resume. Your resume should always be typed or printed on a computer. Design your resume neatly, using an acceptable format (books or your career office can show you some standard formats). Proofread it for errors, and have someone else proofread it as well. Type or print it on a heavier bond paper than is used for ordinary copies. Use only white or off-white paper and black ink.

Interview. Pay attention to your appearance. Be clean, neat, and appropriately dressed. Don't forget to choose a nice pair of shoes—people notice. Bring an extra copy of your resume with you, and any other materials that you want to show the interviewer, even if you have already sent a copy ahead of time. Avoid chewing gum or smoking. Offer a confident handshake. Make eye contact. Show your integrity by speaking honestly about yourself. After the interview is over, no matter what the outcome, send a formal but pleasant thank-you note right away as a follow-up.

Earning the money you need is hard, especially if you work part-time in order to have time for school. Financial aid can take some of the burden off your shoulders. If you can gather one or more loans, grants, or scholarships, they may help make up for what you don't have time to earn.

WHAT SHOULD YOU KNOW ABOUT FINANCIAL AID?

Seeking help from various sources of financial aid has become a way of life for much of the student population. Education is an important but often expensive investment. The cost for a year's full-time tuition only (not including room and board) in 1995–96 ranged from $900 to $15,000, with the national average hovering around $2,100 for public institutions and over $11,000 for private ones.[2] Not many people can pay for tuition in full without aid. In fact, according to the data, over 41 percent of students enrolled received some kind of aid,[3] and that number almost certainly continues to increase along with rising tuition costs.

Most sources of financial aid don't seek out recipients. Take the initiative to learn how you (or you and your parents, if they currently help to support you) can finance your education. Find the people on campus who can help you with your finances. Do some research to find out what's available, weigh the pros and cons of each option, and decide what would work best for you. Try to apply as early as you can. The types of financial aid available to you are loans, grants, scholarships, and professional organizations.

Student Loans

A loan is given to you by a person, bank, or other lending agency, usually to put toward a specific purchase. You, as the recipient of the loan, then must pay back the amount of the loan, plus interest, in regular payments that stretch over a particular period of time. Interest is the fee that you pay for the privilege of using money that belongs to someone else.

Making a Loan Application

What happens when you apply for a loan?

1. *The loaning agency must approve you.* You may be asked about what you and any other family members earn, how much savings you have, your credit history, anything you own that is of substantial value (a home or business), and your history of payment on any previous loans.
2. *An interest charge will be set.* Interest can range from 5 percent to over 20 percent, depending on the loan and the economy. Variable-interest loans shift charges as the economy strengthens or weakens. Fixed-rate loans have one interest rate that remains constant.
3. *The loaning agency will establish a payment plan.* Most loan payments are made monthly or quarterly (four times per year). The payment amount depends on the total amount of the loan, how much you can comfortably pay per month, and the length of the repayment period.

Types of Student Loans

The federal government administers or oversees most student loans. To receive aid from any federal program, you must be a citizen or eligible noncitizen and be enrolled in a program of study that the government has determined is eligible. Individual states may differ in their aid programs. Check with your campus financial aid office to find out details about your state and your school in particular.

Following are the main student loan programs to which you can apply if you are eligible. Amounts vary according to individual circumstances. Contact your school or federal student aid office for further information. In most cases, the amount is limited to the cost of your education minus any other financial aid you are receiving. All the information on federal loans and grants comes from the *1998–1999 Student Guide to Financial Aid*, published by the U. S. Department of Education.[4]

Perkins Loans. Carrying a low, fixed rate of interest, these loans are available to those with exceptional financial need (need is determined by a government-determined formula that indicates how large a contribution toward your education your family should be able to make). Schools issue these loans from their own allotment of federal education funds. After you graduate, you have a "grace period" (up to nine months, depending on whether you were a part-time or full-time student) before you have to begin repaying your loan in monthly installments.

Stafford Loans. Students enrolled in school at least half-time may apply for a Stafford Loan. Exceptional need is not required. However, students who can prove exceptional need may qualify for a "subsidized" Stafford Loan, for which the government pays your interest until you begin repayment. There are two types of Stafford Loans. A Direct Stafford Loan comes from government funds, and an FFEL Stafford Loan comes from a bank or credit union participating in the FFEL (Federal Family Education Loan) program. The type available to you depends on your school's financial aid program. You begin to repay a Stafford Loan six months after you graduate, leave school, or drop below half-time enrollment.

Plus Loans. Your parents can apply for a Plus Loan if they currently claim you as a dependent and if you are enrolled at least half-time. They must also undergo a credit check in order to be eligible, although the loans are not based on income. If they do not pass the credit check, they may be able to sponsor the loan through a relative or friend who does pass. Interest is variable; the loans are available from either the government or banks and credit unions. Your parents will have to begin repayment sixty days after they receive the last loan payment; there is no grace period.

Grants and Scholarships

Both grants and scholarships require no repayment, and therefore give your finances a terrific boost. Grants, funded by the government, are awarded to students who show financial need. Scholarships are awarded to students who show talent or ability in the area specified by the scholarship. They may be financed by government or private organizations, schools, or individuals.

Federal Grant Programs

Pell Grants. These grants are need-based. The Department of Education uses a standard formula to evaluate the financial information you report on your application, and determines your eligibility from that "score" (called an EFC, or Expected Family Contribution, number). You must also be an undergraduate student who has earned no other degrees to be eligible. The Pell Grant serves as a foundation of aid to which you may add other aid sources, and the amount of the grant varies according to the cost of your education and your EFC. Pell Grants require no repayment.

FSEOG (Federal Supplemental Educational Opportunity Grants). Administered by the financial aid administrator at participating schools, FSEOG eligibility depends on need. Whereas the government guarantees that every student eligible for a Pell Grant will receive one, each school receives a limited amount of federal funds for FSEOGs, and once it's gone, it's gone. Schools set their own application deadlines. Apply early. No repayment is required.

Work-Study Program. Although you work in exchange for the aid, work-study is considered a grant because a limited number of positions are available. This program is need-based and encourages community service work or work related in some way to your course of study. You will earn at least the federal minimum wage and will be paid hourly. Jobs can be on-campus (usually for your school) or off-campus (often with a nonprofit organization or a local, state, or federal public agency). Find out who is in charge of the work-study program on your campus.

There is much more to say about these financial aid opportunities than can be touched on here. Many other important details about federal grants and loans are available in The 1998–99 *Student Guide to Financial Aid*. You might find this information at your school's financial aid office, or you can request it by mail, phone, or on-line service:

Address: Federal Student Aid Information Center
                            P. O. Box 84
                            Washington, D. C. 20044

Phone: 1-800-4-FED-AID (1-800-433-3243)
                            TDD for the hearing-impaired: 1-800-730-8913

Internet address: www.ed.gov/prog_info/SFA/StudentGuide

Scholarships

Scholarships are given for different kinds of abilities and talents. Some reward academic achievement. Some reward exceptional abilities in sports or the arts. Some reward citizenship or leadership. Certain scholarships are sponsored by federal agencies. If you display exceptional ability and are disabled, female, of an ethnic background classified as culturally diverse (such as African-American or American Indian), or a child of someone who draws benefits from a state agency (such as a POW or MIA), you might find scholarship opportunities geared toward you.

All kinds of organizations offer scholarships. You may receive scholarships from individual departments at your school or your school's independent scholarship funds, local organizations such as the Rotary Club, or privately operated aid foundations. Labor unions and companies may offer scholarship opportunities for children of their employees. Membership groups such as scouting organizations or the Y might offer scholarships, and religious organizations such as the Knights of Columbus or the Council of Jewish Federations might be another source.

Researching Grants and Scholarships

It can take work to locate scholarships and work-study programs because many of them aren't widely advertised. Ask at your school's financial aid office. Visit your library or bookstore and look in the sections on College or Financial Aid. Guides to funding sources, such as Richard Black's *The Complete Family Guide to College Financial Aid*, catalog thousands of organizations and help you find what fits you. Check out on-line scholarship search services. Use common sense and time management when applying for aid—fill out the application as neatly as possible and send it in on time or even early. In addition, be wary of scholarship scam artists who ask you to pay a fee up front for them to find aid for you.

No matter where your money comes from—financial aid or paychecks from one or more jobs—you can take steps to help it stretch as far as it can go. The next sections concentrate on developing a philosophy about your money and budgeting effectively. Using those skills, you can more efficiently cover your expenses and still have some left over for savings and fun.

How Can Strategic Planning Help You Manage Money?

So you work hard to earn your wages and study hard to hold on to your grants and loans. What do you do with that money? Popular culture tells you to buy. You are surrounded by commercials, magazine ads, and notices in the mail that tell you how wonderful you'll feel if you indulge in some serious spending. On the other hand, there are some definite advantages to not taking that advice. Making some short-term sacrifices in order to save money can help you a great deal in the long run.

Short-Term Sacrifices Can Create Long-Term Gains

When you think about your money, take your values and your ability to plan strategically into account. Ask yourself what goals you value most and what steps you will have to take over time to achieve those goals. You are already planning ahead by being in school and committing to paying for tuition. You may be scrimping now, but you are planning for a career that may reward you with job security and financial stability. Sometimes the most important goals are also the ones that require a long-term commitment. If you can make that commitment, the reward will be worth the short-term sacrifices.

Table 11-3 shows some potential effects of spending. Some effects are negative, some positive, and some more positive than others. Evaluate which you would prefer in the long run. You may find that the pleasure luxuries provide isn't worth the stress created by debt.

"It is thrifty to prepare today for the wants of tomorrow."
AESOP

Critical thinking is the key to smart money planning. Impulsive spending usually happens when you don't take time to think through your decision before you buy. To use your hard-earned money to your greatest benefit, take time to think critically about your finances. First, establish your needs, and be honest about what you truly need and what you just want. Second, brainstorm available options of what to do with your money; evaluate the positive and negative effects of each. Third, choose an option and carry it out. Finally, evaluate the result.

Develop a Financial Philosophy

You can develop your own personal philosophy about spending, saving, and planning. Following are a couple of strategies that you might want to incorporate into that philosophy.

Live beneath your means. Spend less than you make. This strategy helps you create savings, no matter how much or how little. Any amount of savings will give you a buffer zone that can help with emergencies or bigger expenditures. Sometimes your basic needs will cost more than you make, in which case living beneath your means becomes very difficult. If you find, however, that extras are putting your spending over your earnings, cut back.

Table 11-3 Potential effects of spending.

OPTION	POTENTIAL SHORT-TERM EFFECTS	POTENTIAL LONG-TERM EFFECTS
Purchase new sound system	High-quality sound	If paid on credit, a credit card debt, with finance charges, that requires monthly payment; if paid in cash, a loss of benefits that could have come from saving that money
Reduce or pay off credit card debt	Less money for day-to-day expenses; reduction of monthly bills	Improved credit rating and credit history; increased ability to be approved for loans and mortgages; less money charged in interest and fees
Take a week's vacation	Fun and relaxation; stress reduction	Credit card debt or less money saved for future needs
Invest in mutual fund	Less money on hand; more money earning interest	More money earned, due to an interest rate higher than banks can offer
Buy a car	Transportation and independence; gas, maintenance, parking charges	Debt in the form of a car loan; monthly payments for a few years; gradual decrease in car value
Pay health insurance bills	Health insurance coverage; a tighter monthly budget	The safety and security of knowing that your health and the health of your family are protected
Invest in your family business	Commitment to family; less money to spend on extras	Involvement in a family business that can earn you money and provide solid employment for you and other family members
Put money toward tuition	Having to scrimp while in school due to less money on hand; fewer loans and debts	Less money to pay off later in student loans, which means less money charged in interest; more freedom to spend your money on getting settled after you graduate; shorter period of debt

Pay yourself. After you pay your monthly bills, put whatever you can save from your monthly earnings in a savings account. Paying yourself helps you store money in your savings where it can grow.

HOW CAN YOU CREATE A BUDGET THAT WORKS?

Every time you have some money in your pocket and have to figure out whether it will pay for what you want at that moment, you are budgeting your money. It takes some thought and energy to budget efficiently. The more

TERMS

Budgeting
Making a plan for the coordination of resources and expenditures; setting goals with regards to money.

money you can save each month, the more you will thank yourself later when you need it. Consider your resources (money coming in) and expenditures (money flowing out). A smart budget adjusts the money flow for the best possible chance that what comes in will be more than what goes out. Smart budgeting is a worthwhile investment in your future.

The Art of Budgeting

Budgeting involves following a few basic steps in order. These steps are: determining how much money you make, determining how much money you spend, subtracting the second number (what you spend) from the first number (what you make), evaluating the result, and making decisions about how to adjust your spending or earning based on that result. Budgeting regularly is easiest. Use a specified time frame, such as a week or month. Most people budget on a month-by-month basis.

Determine How Much You Will Make

Do this by adding up all your money receipts from the month. If you currently have a regular full-time or part-time job, add your pay stubs. If you have received any financial aid, loan funding, or scholarship money, determine how much of that you can allow for each month's income and add it to your total. For example, if you received a $1,200 grant for the year, each month would have an income of $100. Be sure when you are figuring your income, to use the amounts that remain *after* taxes have been taken out.

Figure Out How Much You Spend

You may or may not have a handle on your spending. Many people don't take the time to keep track. If you have never before paid much attention to how you spend money, examine your spending patterns. Over a month's time, record expenditures in a small notebook or on a piece of paper on a home bulletin board. You don't have to list everything down to the penny. Just indicate expenditures over five dollars, making sure to count smaller expenditures if they are frequent (a bus pass for a month, soda or newspaper purchases per week). In your list, include an estimate of the following:

- Rent/mortgage/school room fees
- Tuition or educational loan payments (divide your annual total by 12 to arrive at a monthly figure)
- Books, lab fees, and other educational expenses
- Regular bills (heat, gas, electric, phone, car payment, water)
- Credit card or other payments on credit
- Food, clothing, toiletries, and household supplies
- Child care
- Entertainment and related items (eating out, books and publications, movies)
- Health, auto, and home/renters' insurance

◆ Transportation and auto expenses

Subtract what you spend from what you make. Ideally, you will have a positive number. You may end up with a negative number, however, especially if you haven't made a habit of keeping track of your spending. This indicates that you are spending more than you make, which over a long period of time can create a nasty debt.

Evaluate the Result

After you arrive at your number, determine what it tells you. If you have a positive number, decide how to save it if you can. If you end up with a negative number, ask yourself questions about what is causing the deficit—where you are spending too much or earning too little. Of course, surprise expenses during some months may cause you to spend more than usual, such as if you have to replace your refrigerator, pay equipment fees for a particular course, or have an emergency medical procedure. However, when a negative number comes up for what seems to be a typical month, you may need to adjust your budget over the long term.

Make Decisions About How to Adjust Spending or Earning

Looking at what may cause you to overspend, brainstorm possible solutions that address those causes. Solutions can involve either increasing resources or decreasing spending. To deal with spending, prioritize your expenditures and trim the ones you really don't need to make. Do you eat out too much? Can you live without cable, a beeper, a cellular phone? Be smart. Cut out unaffordable extras. As for resources, investigate ways to take in more money. Taking a part-time job, hunting down scholarships or grants, or increasing hours at a current job may help.

A Sample Budget

Table 11-4 shows a sample budget of an unmarried student living with two other students. It will give you an idea of how to budget (all expenditures are general estimates, based on averages).

To make up the $190 that this student went over budget, he can adjust his spending. He could rent movies or check them out of the library instead of going to the theater. He could socialize with friends at someone's apartment instead of paying high prices and tips at a bar or restaurant. Instead of buying CDs and tapes, he could borrow them. He could also shop for specials and bargains in the grocery store or go to a warehouse supermarket to stock up on staples at discount prices. He could make his lunch instead of buying it and walk instead of taking public transportation.

Not everyone likes the work involved in keeping a budget. While linear, factual, reflective, and verbal learners may take to it more easily, active, holistic, theoretical, and visual learners may resist the structure and detail (see Chapter 3). Visual learners may want to create a budget chart like the one shown in the example or construct a think link that shows the connections between all the month's expenditures. Use images to clarify ideas, such as pic-

Table 11-4 — A student's sample budget.	Part-time salary: $10 an hour, 20 hours a week. 10 × 20 = $200 a week, × 4 1/3 weeks (one month) = $866. Student loan from school's financial aid office: $2,000 divided by 12 months = $166. Total income per month: $1,032.

MONTHLY EXPENDITURES	AMOUNT
Tuition ($6500 per year)	$ 542
Public transportation	$ 90
Phone	$ 40
Food	$ 130
Medical insurance	$ 120
Rent (including utilities)	$ 200
Entertainment/miscellaneous	$ 100
Total spending	$1222

$1032 (income) − $1222 (spending) = $−190 ($190 over budget)

turing a bathtub you are filling that is draining at the same time. Use strategies that make budgeting more tangible, such as dumping all of your receipts into a big jar and tallying them at the end of the month. Even if you have to force yourself to do it, you will discover that budgeting can reduce stress and help you take control of your finances and your life.

Savings Strategies

You can save money and still enjoy life. Make your fun less expensive and environmentally friendly—or save up for a while to splurge on a really special occasion. Here are some suggestions for saving a little bit of money here and there. Small amounts can add up to big savings after a while.

- Attend bargain movies.
- When safe for the fabric, hand-wash items you ordinarily dry-clean, or don't buy items that need dry-cleaning in the first place.
- Check movies, CDs, tapes, and books out of your library.
- Make popcorn instead of buying bags of chips.
- Walk or bike instead of paying for public transportation or driving.
- If you have storage space, buy detergent, paper products, toiletries, and other staples in bulk.
- Shop in secondhand stores.
- Keep your possessions neat, clean, and properly maintained—they will last longer.
- Take advantage of weekly supermarket specials, and bring coupons when you shop.
- Reuse grocery bags for food storage and garbage instead of buying bags.
- Return bottles and cans for deposits if you live in a state that accepts them.

- Trade clothing with friends and barter services (plumbing for babysitting, for example).
- Buy display models of appliances or electronics (stereo equipment, TVs, VCRs).
- Take your lunch instead of buying it.
- Find a low-rate long-distance calling plan, use e-mail, or write letters.
- Save on heat by dressing warmly and using blankets; save on air conditioning by using fans.
- Have potluck parties; ask people to bring dinner foods or munchies.

Add your own suggestions here!

___

___

___

You can also maximize savings and minimize spending by using credit cards wisely.

Managing Credit Cards

Most credit comes in the form of a powerful little plastic card. Credit card companies often solicit students on campus or through the mail. When choosing a card, pay attention to the *annual fee* and *interest rates,* the two ways in which a credit card company makes money from you. Some cards have no annual fee; others may charge a flat rate of $10 to $70 per year. Interest rates can be fixed or variable. A variable rate of 12 percent may shoot up to 18 percent when the economy slows down. You might be better off with a mid-range fixed rate that will always stay the same.

Following are some potential effects of using credit.

Positive Effects

Establishing a good credit history. If you use your credit card moderately and pay your bills on time, you will make a positive impression on your creditors. Your *credit history* (the record of your credit use, including positive actions such as paying on time and negative actions such as going over your credit limit) and *credit rating* (the score you are given based on your history) can make or break your ability to take out a loan or mortgage. How promptly you make loan payments and pay mortgage and utility bills affects your credit rating as well. Certain companies track your credit history and give you a credit rating. Banks or potential employers will contact these companies to see if you are a good credit risk.

Creditors
People to whom debts are owed, usually money.

Emergencies. Few people carry enough cash to handle unexpected expenses. Your credit card can help you in emergency situations such as when your car needs to be towed or your bike gets a flat.

Record of purchases. Credit card statements give you a monthly record of purchases made, where they were made, and exactly how much was paid. Using your credit card for purchases that you want to track, such as work expenses, can help you keep records for tax purposes.

Negative Effects

Credit can be addictive. Credit can be like a drug, seeming fun because the pain of paying is put off until later. If you get hooked, though, you can wind up thousands of dollars in debt to creditors. The high interest will enlarge your debt; your credit rating may fall, potentially hurting your eligibility for loans and mortgages; and you may lose your credit cards altogether.

Credit spending can be hard to monitor. Paying by credit can seem so easy that you don't realize how much you are spending. When the bill comes at the end of the month, the total can hit you hard.

You are taking out a high-interest loan. Buying on credit is similar to taking out a loan—you are using money with the promise to pay it back. Loan rates, however, especially on fixed-interest loans, are often much lower than the 11 to 23 percent on credit card debt. Fifteen percent interest per year on a credit card debt averaging $2000 is approximately $300; 5 percent interest per year on a loan in the same amount is $100.

Bad credit ratings can haunt you. Any time you are late with a payment, default on a payment, or in any way misuse your card, a record of that occurrence will be entered on your credit history, lowering your credit rating. If a prospective employer or loan officer discovers a low rating, you will seem less trustworthy and may lose the chance at a job or a loan.

Managing Credit Card Debt

"Put not your trust in money, but put your money in trust."

OLIVER WENDELL HOLMES

There are ways to manage credit card debt so that it doesn't get worse. Stay in control by having only one or two cards and paying bills regularly and on time. Try to pay in full each month. If you can't, at least pay the minimum. Make as much of a dent in the bill as you can.

If you get into trouble, three steps will help you deal with the situation. First, *admit* that you made a mistake, even though you may be embarrassed. Then, *address* the problem immediately and honestly in order to minimize the damages. Call the bank or credit card company to talk to someone about the problem. They may draw up a payment plan that allows you to pay your debt gradually, in amounts that your budget can manage. Creditors would rather accept small payments than nothing at all.

Finally, *prevent* this problem from happening again. Figure out what got you into trouble and take steps to avoid it in the future if you can. Some financial disasters, such as medical emergencies, may be beyond your control. Overspending on luxuries, however, is something you have the power to avoid. Make a habit of balancing your checkbook. Cut up a credit card or two if you have too many. Don't let a high credit limit tempt you to spend. Pay every month, even if you pay only the minimum. If you work to clean up your act, your credit history will gradually clean up as well.

REAL WORLD PERSPECTIVE

What should I do about all these credit card offers?

Brett Cross, University of Washington

I am a pre-engineering student at the University of Washington. Recently, I have been receiving a number of credit card applications offering a low interest rate. In fact, I get at least one offer a week. I've been thinking it would be nice to establish credit, but I'm not sure if getting a credit card right now is a good idea. Even though I have a part-time job and have financial aid, it seems like there's never enough to make it to the end of the semester. Should I apply for one of these credit cards? It would be really great to have some extra cash every now and then.

Tim Short, Washington State University

Dealing with financial hardships while in college is a part of life for many people these days. Credit card offers are in abundance for college students, and for good reason. Credit companies know that most college students won't be able to pay off their cards until after they graduate, and that they tend to carry balances and pay interest and hefty fees until they are solvent. Believe me, I know. Throughout my past four years at college, I have acquired several credit cards. On them I have charged things such as books, car repairs, auto insurance, and other personal items. I am still paying interest on these cards monthly and will not be able to pay them off until after I graduate.

My suggestion to you is this: Don't take out a credit card unless you absolutely have to. If you can take out student loans or borrow from your parents, do that instead. Most academic loans have a 6 percent to 8 percent interest rate, which is much lower than the 18 percent to 21 percent that most credit card companies charge. Don't be fooled by offers for a card with a low rate. These invariably expire after one year and then the rate jumps up. If you miss a payment during that year, some companies will raise your rates immediately. Rationalizing that you will pay the card off before that time frame is up is also not a good idea. Unless you are on the verge of graduation, you will probably not have any more cash in a year than you do now. Overall, my advice is this: If you can avoid borrowing from credit card companies, do so! You will be a lot happier in the long run.

Sacrifici

In Italy, parents often use the term *sacrifici*, meaning "sacrifices," to refer to tough choices that they make in order to improve the lives of their children and family members. They may sacrifice a larger home so that they can afford to pay for their children's sports and after-school activities. They may sacrifice a higher-paying job so that they can live close to where they work. They give up something in exchange for something else that they have decided is more important to them.

Think of the concept of *sacrifici* as you analyze the sacrifices you can make in order to get out of debt, reach your savings goals, and prepare for a career that you find satisfying. Many of the short-term sacrifices you are making today will help you do and have what you want in the future. Keep that notion as a light to guide you through the ups and downs of student life.

Chapter 11 Applications

Name _____ Date _____

KEY INTO YOUR LIFE
Opportunities to Apply What You Learn

11.1 Mentors

First, consider the people you go to with problems and questions, people whom you trust and with whom you share a lot of yourself. Name up to three—don't fill the list unless you can really think of three people you trust.

1. _____
2. _____
3. _____

Evaluate your list. With which of those people do you feel you could have a mentoring relationship? Name up to two; for each, name two steps you can take to invest even further in your relationship.

1. _____
_____
2. _____
_____

11.2 Your Job Priorities

What kind of a job could you manage while you're in school? How would you want a job to benefit you? Discuss your requirements in each of the following areas.

Salary/wage level _____
Time of day _____
Hours per week (part-time vs. full-time) _____
Duties _____
Location _____

Flexibility _____

Affiliation with school or financial aid program _____

- What kind of job might fit all or most of your requirements? List two possibilities here.

    1. _____
    2. _____

 ## Savings Brainstorm

As a class, brainstorm areas that require financial management (such as funding an education, running a household, or putting savings away for the future) and write them on the board. Divide into small groups. Each group should choose one area to discuss (make sure all areas are chosen). In your group, brainstorm strategies that can help with the area you have chosen. Think of savings ideas, ways to control spending, ways to earn more money, and any other methods of relieving financial stress. Agree on a list of possible ideas for your area, and share it with the class.

KEY TO SELF-EXPRESSION
Discovery Through Journal Writing

To record your thoughts, use a separate journal or the lined page at the end of the chapter.

Credit Cards

Describe how you use credit cards. Are you conservative, overindulgent, or in between? Do you pay on time, and do you pay the full balance of the card, or not? How does using a credit card make you feel—powerful, excited, apprehensive, or nervous? For what sort of purchases do you use credit cards? If you would like to change how you use credit, discuss any changes you want to make and how they would help you.

Journal

Name _____ Date _____

12 Moving Ahead

Building a Smart Future

The end of one path can be the beginning of another. For example, graduation is often referred to as commencement, because the end of your student career is the beginning or renewal of your life as a working citizen. As you come to the end of your work in this course, you have built up a wealth of knowledge. Now you have more power to make decisions about what directions you want your studies, your career, and your personal growth to take.

This chapter will explore how to manage the constant change you will encounter. Developing your flexibility will enable you to adjust goals, make the most of successes, and work through failures. You will consider what is important about giving back to your community and continuing to learn throughout your life. Finally, you will revisit your personal mission, exploring how to revise it as you encounter changes in the future.

In this chapter, you will explore answers to the following questions:

◆ What are some of the big questions in science today?

- How can you live with change?
- What will help you handle success and failure?
- Why give back to the community and the world?
- Why is college just the beginning of lifelong learning?
- How can you live your mission?

WHAT ARE SOME OF THE BIG QUESTIONS IN SCIENCE TODAY?

Biotechnology and genetics are examples of the rapid changes occurring in science today. The Human Genome Project, an international effort launched in 1989, plans to map the entire human genome by the year 2005.[1] But genetic innovations have been used in health care for years; examples include the production of insulin, human hemoglobin produced in pigs, and Factor IX for hemophilia in sheep's milk. Newer innovations include genetic disease treatment, or gene therapy, which places a fully functioning gene into cells to replace, or augment, the function of a defective gene. At this time gene therapy is primarily experimental, but that will soon change as techniques are improved and tested.

Questions about the use of new technology and discoveries arise in all areas of science. Genetics is a good example of how questions concern not only scientists but citizens as well. For instance, gene therapy that affects only somatic cells, body cells that are not involved in reproduction, will not affect future generations. On the other hand, gene therapy performed on germ cells, the cells of reproduction, alters the genes so that these changes are passed on to future generations. This raises many important questions concerning the desirability of permanently altering the human gene pool. Most geneticists currently agree that germ cell therapy is not advisable.[2]

More recently, the use of stem cells from non-viable fetuses has been discussed. These cells have the possibility of regenerating human tissue. For example, experiments are being done with stem cells to see if they could be used in humans to grow arterial bypasses in the heart. If this works, many cardiac surgeries and invasive procedures would become unnecessary. This potential life- and cost-saving therapy raises ethical concerns for some people. Implications of scientific research must be understood by scientists and non-scientists or potentially breakthrough work may be overlooked and underfunded due to decisions based on uninformed reactions. Likewise, ethical issues must be equally considered. Continuing with genetics as an example, some of the factors to consider include the legal, economic, social, and ethical implications.

Legal Implications

The *Yale Law Journal* states that because knowledge of human genetics is expanding so rapidly, keeping current with its growth is imperative.[3] Understanding is important for everyone because advances in gene mapping will allow genetic testing that can predict predisposition to diseases such as heart disease, obesity, cancer, and diabetes. As these tests become available, questions arise such as: Who gets tested? Are the tests accurate? Who sees the test results? Concerns about privacy and confidentiality affect everyone.

Economic Implications

Genetics studies and testing can be expensive. The March of Dimes estimates that 3 percent to 5 percent of all live births have some form of genetic abnormality and that there are close to 40,000 genetic disorders. If a national screening program was implemented for cystic fibrosis alone, the costs would be approximately $2.2 million for each case that was avoided.[4]

Funding projections for Human Genome Project research are $200 million each year for the next ten to fifteen years. Approximately 5 percent of this is budgeted for the consideration of ethical, social, and legal issues.[5] Questions arise such as: Is it fair to spend vast sums on genetic research when so many go without basic health care? Are the potential benefits of genetic research, such as preventing and eventually treating gene disorders, worth the cost? On the other hand, and of equal concern, is the vast benefit of understanding, detecting, and treating genetic disorders and the potential savings by avoiding costly treatments as a disease progresses.

Social Implications

The social concerns raised by genetics are broad, and the ability to view genetic research with a critical eye is crucial to understanding them. For example, the media often sensationalize genetics. An article in the *Los Angeles Times* claimed that DNA predicts whether we will be "hard-driving executives or moon-bedazzled poets."[6] Questions about gender selection of fetuses by potential parents, attributing learned behaviors to genes, and having knowledge that others carry predispositions to diseases are being considered by geneticists, those in the health sciences, and by the public with serious concern.[7]

Ethical Implications

As a scientist you must be familiar with ethical dilemmas. For instance, in genetics the ability to test for the predisposition to diseases may pose a risk to confidentiality and privacy. Discrimination based on genetic test results could be grounds for denial of employment or insurance, although the Americans with Disabilities Act may offer protection. People will need to be educated on test results and the possible consequences of releasing them.

The ability to perform gene therapy raises many ethical questions. "Disorder," "defect," "error," and "mutation" are words we often use when discussing genetic variations. They clearly imply failure. Will we become legally or morally bound to fix everything with gene therapy?

Genetics, along with many other areas of research, offers great opportunities to learn more about human physiology, disease, and the world around us. But this new knowledge must be thought about critically. It is vitally important that you, as a science major, take at least one ethics course, especially if you are planning on a career in the health sciences. Technology is advancing so quickly that it can outpace our ability to consider fully the implications of putting it into practice.

Something to Prove: Does the Truth Exist?

Ask yourself these questions: Is there an absolute truth? Who decides what is true and what is not true? Remember when you consider these questions that some people think there is evidence that the position of the stars at the time of their birth determines their future; others that the Christian Bible holds literal truths; and still others believe that women and girls do not need an education. How can you decide on the perplexing issues in science? Can you decide what is the right thing for everyone?

Science Is Multidisciplinary

Thinking about these tough questions, and others like them, will help you understand how science is also a philosophical, religious, social, and political pursuit. The more you understand these areas, along with science, the better off you'll be in planning and making decisions that affect you, your family, and your local and global communities. A thorough background in the sciences and in the liberal arts is a necessity in science and will help you in any career you choose. Big questions about truth and decisions based on values occur everywhere and they will occur throughout your lifetime.

Scientific Discoveries and Truth

Scientific discovery has been described as similar to peeling an onion. This analogy is based on the premise that there exists absolute truth and that each discovery removes a layer of the onion, bringing us one step closer to this truth. The assumption that a single truth exists is questioned by scientists who view scientific discovery not as seeking or finding an absolute truth but as adding to the body of knowledge about a subject. This body of knowledge then lends support to certain ideas, or hypotheses, creating theories. The more evidence there is for a theory, the stronger that theory is. Rarely are the words *truth* or *proof* used in the sciences except in the context of philosophical or personal views and values.

> "Nothing causes as much destruction, misery, and death as obsession with a truth believed to be absolute. Every crime in history is the product of some fanaticism. Every massacre is performed in the name of virtue; in the name of legitimate nationalism, a true religion, a just ideology, the fight against Satan."
>
> FRANÇOIS JACOB

How can you live with change?

Even the most carefully constructed plans can be turned upside down by change. In this section, you will explore some ways to make change a manageable part of your life by accepting the reality of change, maintaining flexibility, and adjusting your goals.

Accept the Reality of Change

As Russian-born author Isaac Asimov once said, "It is change, continuing change, inevitable change, that is the dominant factor in society today. No sensible decision can be made any longer without taking into account not only the world as it is, but the world as it will be."[8] Change is a sure thing. Two significant causes of change on a global level are technology and the economy.

Technological Growth

Today's technology has spurred change. Tasks that people have performed for years are now taken care of by computer in a fraction of the time and for a fraction of the price. Advances in technology come into being daily: Computer companies update programs, new models of cars and machines appear, and scientists discover new possibilities in medicine and other areas. People make changes in the workplace, school, and home to keep up with the new systems and products that technology constantly offers. People and cultures are linked around the world through the Internet and World Wide Web.

The dominance of the media, brought on by technological growth, has increased the likelihood of change. A few hundred years ago, no television or magazines or Internet existed to show people what was happening elsewhere in the world. A village could operate in the same way for years with very little change, because there would be little to no contact with anyone from the outside who could introduce new ideas, methods, or plans. Now, the media constantly presents people with new ways of doing things. When people can see the possibilities around them, they are more likely to want to find out whether the grass is truly greener on the other side of the fence.

Economic Instability

The unpredictable economy is the second factor in this age of constant change. Businesses have had to cut costs in order to survive, which has affected many people's jobs and careers. Some businesses discovered the speed and cost-effectiveness of computers and used them to replace workers. Some businesses have had to downsize and have laid off people to save money. Some businesses have merged with others, and people in duplicate jobs were let go. The difficult economy has also had an effect on personal finances. Many people face money problems at home that force them to make changes in how much they work, how they pursue an education, and how they live.

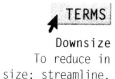
Downsize
To reduce in size; streamline.

Maintain Flexibility

The fear of change is as inevitable as change itself. When you become comfortable with something, you tend to want it to stay the way it is, whether it is a relationship, a place you live, a job, a schedule, or the racial/cultural mix of people with whom you interact. Change may seem to have only negative effects, and consistency only positive effects. Think about your life right now. What do you wish would always stay the same? What changes have upset you and thrown you off balance?

You may have encountered any number of changes in your life to date, many of them unexpected. You may have experienced ups and downs in relationships, perhaps marriage or divorce. You may have changed schools, changed jobs, or moved to a new home. You may have shifted your course of study. You may have added to your family or lost family members.

Financial shifts may have caused you to change the way you live. All of these changes, whether you perceive them as good or bad, cause a certain level of stress. They also cause a shift in your personal needs, which may lead to changing priorities.

Change Brings Different Needs

Your needs can change from day to day, year to year, and situation to situation. Although you may know about some changes ahead of time, such as when you plan to attend school or move in together with a partner, others may take you completely by surprise, such as losing a job, illness, or an unexpected pregnancy. Even the different times of year bring different needs, such as a need for extra cash around the holidays or a need for additional child care when your children are home for the summer.

Some changes that shift your needs will occur within a week or even a day. For example, an instructor may inform you that you have a quiz or extra assignment at the end of the week, or your supervisor at work may give you an additional goal for the week. During the course of a day, your daughter might tell you that she needs you to drive her somewhere that evening, or a friend may call and need your help with something that has come up suddenly. Table 12-1, on the following page, shows how the effects of certain changes can lead to new priorities.

Flexibility vs. Inflexibility

When change affects your needs, *flexibility* will help you shift your priorities so that you address those needs. You can react to change with either inflexibility or flexibility, each with its resulting effects.

Inflexibility. Not acknowledging a shift in needs can cause trouble. For example, if you lose your job and continue to spend as much money as you did before, ignoring your need to live more modestly, you can drive yourself into debt and make the situation worse. Or if you continue to spend little time with a partner who has expressed a need for more contact, you may lose your relationship.

Flexibility. Being flexible means acknowledging the change, examining your different needs, and addressing them in any way you can. As frightening as it can be, being flexible can help you move ahead. Discovering what change brings may help you uncover positive effects that you had no idea were there. For example, a painful breakup or divorce can lead you to discover greater capability and independence. A loss of a job can give you a chance to reevaluate your abilities and look for another job in an area that suits you better. An illness can give you perspective on what you truly value in life. In other words, a crisis can spur opportunity; you may learn that you want to adjust your goals in order to pursue that opportunity.

Sometimes you may need to resist for a while, until you are ready to face an important change. When you do decide you are ready, being flexible will help you cope with the negative effects and benefit from the positive effects.

Table 12-1   Change produces new priorities.

CHANGE	EFFECTS AND CHANGED NEEDS	NEW PRIORITIES
Lost job	Loss of income; need for others in your household to contribute more income	Job hunting; reduction in your spending; additional training or education in order to qualify for a different job
New job	Change in daily/weekly schedule; need for increased contribution of household help from others	Time and energy commitment to new job; maintaining confidence; learning new skills
Started school	Fewer hours for work, family, and personal time; responsibility for classwork; need to plan semesters ahead of time	Careful scheduling; making sure you have time to attend class and study adequately; strategic planning of classes and of career goals
Relationship/marriage	Responsibility toward your partner; merging of your schedules and perhaps your finances and belongings	Time and energy commitment to relationship
Breakup/divorce	Change in responsibility for any children; increased responsibility for your own finances; possibly a need to relocate; increased independence	Making time for yourself; gathering support from friends and family; securing your finances; making sure you have your own income
Bought car	Responsibility for monthly payment; responsibility for upkeep	Regular income so that you can make payments on time; time and money for upkeep
New baby	Increased parenting responsibility; need money to pay for baby or if you had to stop working; need help with other children	Child care; flexible employment; increased commitment from a partner or other supporter
New cultural environment (from new home, job, or school)	Exposure to unfamiliar people and traditions; tendency to keep to yourself	Learning about the culture with which you are now interacting; openness to new relationships

Adjust Your Goals

Your changing life will often result in the need to adjust goals accordingly. Sometimes goals must change because they weren't appropriate in the first place. Some turn out to be unreachable; some may not pose enough of a challenge; others may be unhealthy for the person who set them or harmful to others.

Step One: Reevaluate

Before making adjustments in response to change, take time to *reevaluate* both your goals and your progress toward them.

Your goals. First, determine whether your goals still fit the person you have become in the past week, month, or year. Circumstances can change quickly.

For example, an unexpected pregnancy might cause a female student to rethink her educational goals.

Your progress. If you feel you haven't gotten far, determine whether the goal is out of your range or simply requires more stamina than you had anticipated. As you work toward any goal, you will experience alternating periods of progress and stagnation. Sticking with a tough goal may be the hardest thing you'll ever do, but the payoff may be worth it. Seek the support and perspective of a friend or counselor as you evaluate your progress.

Step Two: Modify

If after your best efforts it becomes clear that a goal is out of reach, *modifying* your goal may bring success. Perhaps the goal doesn't suit you. For example, an active, interpersonal learner might become frustrated while pursuing a detail-oriented, sedentary career such as computer engineering.

Based on your reevaluation, you can modify a goal in two ways:

1. Adjust the existing goal. To adjust a goal, change one or more aspects that define that goal—for example, the time frame, the due dates, or the specifics of the expectations. For example, a woman with an unexpected pregnancy could adjust her educational due date, taking an extra year or two to complete her course work. She could also adjust the time frame, taking classes at night if she had to care for her child during the day.
2. Replace it with a different, more compatible goal. If you find that you just can't handle a particular goal, try to find another that makes more sense for you at this time. For example, a couple who wants to buy a home but just can't afford it can choose to work toward the goal of making improvements to their current living space. Because you and your circumstances never stop changing, your goals should keep up with those changes.

Being open to adjusting your goals will help you manage both failure and success along the way.

WHAT WILL HELP YOU HANDLE SUCCESS AND FAILURE?

The perfect, trouble-free life is only a myth. The most wonderful, challenging, fulfilling life is full of problems to be solved and difficult decisions to be made. If you want to handle the bumps and bruises without losing your self-esteem, you should prepare to encounter setbacks along with your successes.

Dealing With Failure

Things don't always go the way you want them to go. Sometimes you may come up against obstacles that are difficult to overcome. Sometimes you will let yourself down or disappoint others. You may make mistakes or lose your motivation. All people do, no matter who they are or how smart or

> "Risk! Risk anything! Care no more for the opinion of others, for those voices. Do the hardest thing on earth for you. Act for yourself. Face the truth."
>
> KATHERINE MANSFIELD

accomplished they may be. What is important is how you choose to deal with what goes wrong. If you can arrive at reasonable definitions of failure and success, accept failure as part of being human, and examine failure so that you can learn from it, you will have the confidence to pick yourself up and keep improving.

Measuring Failure and Success

Most people measure failure by comparing where they are to where they believe they should be. Since individual circumstances vary widely, so do definitions of failure. What you consider a failure may seem like a positive step for someone else. Here are some examples:

- Imagine that your native language is Spanish. You have learned to speak English well, but you still have trouble writing it. Making writing mistakes may seem like failure to you, but to a recent immigrant from the Dominican Republic who knows limited English, your command of the language will seem like a success story.
- If two people apply for internships, one may see failure as receiving some offers but not the favorite one, while someone who was turned down may see any offer as a success.
- Having a job that doesn't pay you as much as you want may seem like a failure, but to someone who is having trouble finding any job, your job is a definite success.

Accepting Failure

No one escapes failure, no matter how hard he or she may try (or how successful he or she may be at hiding mistakes). The most successful people and organizations have experienced failures and mistakes. For example, the producers of the film *Waterworld* spent over $140 million on a film that made only a fraction of that cost at the box office. America Online miscalculated customer use and offered a flat rate per month, resulting in thousands of customers' having trouble logging on to the service. Many an otherwise successful individual has had a problematic relationship, a substance abuse problem, or a failing grade in a course.

You have choices when deciding how to view a failure or mistake. You can pretend it never happened, blame it on someone or something else, blame yourself, or forgive yourself.

Pretending it didn't happen. Avoiding the pain of dealing with a failure can deny you valuable lessons and could even create more serious problems. HIV is one example of this idea. Imagine that a person has unprotected sex with a potentially HIV-infected partner and then denies it ever happened. If that person later discovers that he or she has contracted HIV from the first partner, the deadly virus may have been passed on to any subsequent partners.

Blaming others. Putting the responsibility on someone else stifles opportunities to learn and grow. For example, imagine that an unprepared and inappropriately dressed person interviews for a job and is not hired. If he or

"History is what you remember, and if you don't think it's being revised all the time, you haven't paid enough attention to your own memory. When you remember something, you don't remember the thing itself—you just remember the time you remembered it."

JOHN BARLOW

she decides that the interviewer is biased, the interviewee won't learn to improve preparation or interview strategies. Evaluate causes carefully and try not to assign blame.

Blaming yourself. Getting angry at yourself for failing, or believing that you should be perfect, can only result in your feeling incapable of success and perhaps becoming afraid to try. Negative self-talk can become self-fulfilling.

Forgiving yourself. This is by far the best way to cope. First, although you should always strive for your best, don't expect perfection of yourself or anyone else. Expect that you will do the best that you can within the circumstances of your life. Just getting through another day as a student, employee, and/or parent is an important success. Second, forgive yourself when you fail. Your value as a human being does not diminish when you make a mistake. Forgiving yourself will give you more strength to learn from the experience, move on, and try again.

Once you are able to approach failure and mistakes in a productive way, you can explore what you can learn from them.

Learning From Failure

Learning from your failures and mistakes involves thinking critically through what happened. The first step is to evaluate what happened and decide if it was within your control. It could have had nothing to do with you at all. You could have failed to win a job because someone else with equal qualifications was in line for it ahead of you. A family crisis that disrupted your sleep could have affected your studying, resulting in a failing grade on a test. These are unfortunate circumstances, but they are not failures. On the other hand, something you did or didn't do may have contributed to the failure.

If you decide that you have made a mistake, your next steps are to analyze the causes and effects of what happened, make any improvements that you can, and decide how to change your action or approach in the future.

For example, imagine that after a long night of studying, you forgot your part-time work-study commitment the next day.

Analyze causes and effects. *Causes:* Your exhaustion and your concern about your test caused you to forget to check on your work schedule. *Effects:* Because you weren't there, a crucial curriculum project wasn't completed. An entire class and instructor who needed the project have been affected by your mistake.

Make any possible improvements on the situation. Apologize to the instructor and see if there is still a chance to finish up part of the work that day.

Make changes for the future. You could set a goal to note your work schedule regularly in your date book—maybe in a bright color—and to check it more often. You could also arrange your future study schedule so that you will be less exhausted.

Think about the people you consider exceptionally successful. They didn't rise to the top without taking risks and making their share of mistakes. They have built much of their success upon their willingness to recognize and learn from their shortfalls. You too can benefit from staying open to this kind of active, demanding, hard-won education. Learning involves change and growth. Let what you learn from falling short of your goals inspire new and better ideas.

REAL WORLD PERSPECTIVE

How can I prepare to make a difference in the world when I finish college?

Norma Espina, University of Texas—El Paso

Right after high school I tried college and was very unsuccessful at it. I didn't realize what I was getting into. When I was in high school, I was surrounded by my friends. If someone didn't know me personally, they at least knew who I was. College was so different; no one knew me. I wanted to appear grown up so I didn't risk very much. This was one of my downfalls. I was afraid to ask for help or get involved because I wanted to be mature, and I was too afraid to make a mistake. This backfired on me because I started to fall behind in classes. When I didn't understand something, I just let it slide by. Before long, I was avoiding classes. I had excuse after excuse after excuse until finally I didn't want to go to college anymore. That's when I gave up.

Seven years later, after a divorce and two children, I decided to return and finish my education. I was very motivated to succeed. I believe the reason I am successful this time is because I am willing to get involved. I ask questions in class or talk to the professor right after. Through speaking up I began making friends and forming study groups.

My dream is to finish college so I can begin making a difference in the world. I believe I'm on the right track and have a positive attitude about the direction I'm headed. What specific steps do you recommend I take as I prepare for my future outside of college?

Mike Jackson, Baltimore City Community College

Whether you are in school or beginning your career, one of the things that will contribute to your success is to feel positive about who you are and the dreams you have. Sometimes people who enter the job force allow the job or the group they're in to define them. I personally feel it's better to let the group or the job *enhance* who you are, but not control your life. That's why it's also important to have balance between your work and your personal life. The more balanced you are, the greater the chance you'll have a healthier perspective on your job and on people in general.

I also believe that life is a series of opportunities, so when one comes along, you've got to grab onto it. It's important to not let life just happen to you. Otherwise you could find yourself in circumstances which are very unpleasant, to say the least. Growing up in an inner-city environment as I did opens your eyes to what can occur when you let life happen to you. Fortunately for me, I had parents with very strong values to help point me in the right direction. But even without supportive parents like mine, if you believe that your goals are worth having, you can make it out of the worst of circumstances.

Finally, have a plan. Decide what you want to do with your life and then formulate steps to achieve that goal. Have some alternatives, too, in case your original ideas don't pan out. But don't worry if you stray from your original plan. Some detours can actually be better than the goal you had in the first place.

Think Positively About Failure

When you feel you have failed, how can you boost your outlook?

Stay aware of the fact that you are a capable, valuable person. People often react to failure by becoming convinced that they are incapable and incompetent. Fight that tendency by reminding yourself of your successes, focusing your energy on your best abilities, and knowing that you have the strength to try again. Realize that your failure isn't a setback as long as you learn from it and rededicate yourself to excellence. Remember that the energy you might expend on talking down to yourself would be better spent on trying again and moving ahead.

Share your thoughts and disappointment with others. Everybody fails. Trading stories will help you realize you're not alone. People refrain from talking about failures out of embarrassment, often feeling as though no one else could have made as big a mistake as they did. When you open up, though, you may be surprised to hear others exchange stories that rival your own. Be careful not to get caught in a destructive cycle of complaining. Instead, focus on the kind of creative energy that can help you find ways to learn from your failures.

Look on the bright side. At worst, you at least have learned a lesson that will help you avoid similar situations in the future. At best, there may be some positive results of what happened. If your romance flounders, the extra study time you suddenly have may help you boost your grades. If you fail a class, you may discover that you need to focus on a different subject that suits you better. What you learn from a failure may, in an unexpected way, bring you around to where you want to be.

Dealing With Success

Success isn't reserved for the wealthy, famous people you see glamorized in magazines and newspapers. Success isn't money or fame, although it can bring such things. Success is being who you want to be and doing what you want to do. Success is within your reach.

Pay attention to the small things when measuring success. You may not feel successful until you reach an important goal you have set for yourself. However, along the way each step is a success. When you are trying to drop a harmful habit, each time you stay on course is a success. When you are juggling work, school, and personal life, just coping with what every new day brings equals success. If you received a C on a paper and then earned a B on the next one, your advancement is successful.

Remember that success is a process. If you deny yourself the label of "success" until you reach the top of where you want to be, you will have a much harder time getting there. Just moving ahead toward improvement and growth, however fast or slow the movement, equals success.

Here are some techniques to handle your successes.

First, appreciate yourself. You deserve it. Take time to congratulate yourself for a job well done—whether it be a good grade, an important step in learning a new language, a job offer, a promotion or graduation, or a personal victory over substance abuse. Bask in the glow a bit. Everybody hears about his or her mistakes, but people don't praise themselves (or each other) enough when success happens. Praise can give you a terrific vote of confidence.

Take that confidence on the road. This victory can lead to others. Based on this success, you may be expected to prove to yourself and others that you are capable of growth, of continuing your successes and building upon them. Show yourself and others that the confidence is well founded.

Stay sensitive to others. There could be people around you who may not have been so successful. Remember that you have been in their place, and they in yours, and the positions may change many times over in the future. Enjoy what you have, work to build on it and not to take it for granted, and support others as they need it.

Staying sensitive to others is an important goal always, whether you are feeling successful or less than successful. Giving what you can of your time, energy, and resources to the community and the world is part of being aware of what others need. Your contributions can help to bring success to others.

WHY GIVE BACK TO THE COMMUNITY AND THE WORLD?

Everyday life is demanding. You can become so caught up in the issues of your own life that you neglect to look outside your immediate needs. However, from time to time you may feel that your mission extends beyond your personal life. You have spent time in this course working to improve yourself. Now that you've come so far, why not extend some of that energy and effort to the world outside? With all that you have to offer, you have the power to make positive differences in the lives of others. Every effort you make, no matter how small, improves the world.

Your Imprint on the World

As difficult as your life can sometimes seem, looking outside yourself and into the lives of others can help put everything in perspective. Sometimes you can evaluate your own hardships more reasonably when you look at them in light of what is happening elsewhere in the world. There are always many people in the world in great need. You have something to give to others. Making a lasting difference in the lives of others is something to be proud of.

Your perspective may change after volunteering at a soup kitchen. Your appreciation of those close to you may increase after you spend time with cancer patients at the local hospice. Your perspective on your living situation may change after you help people improve their housing conditions.

If you could eavesdrop on someone *talking about you* to another person, what do you think you would hear? How would you like to hear yourself described? What you do for others makes an imprint that can have far more impact than you may imagine. Giving one person hope, comfort, or help can improve his or her ability to cope with life's changes. That person in turn may be able to offer help to someone else. As each person makes a contribution, a cycle of positive effects is generated. For example, Helen Keller, blind and deaf from the age of 2, was educated through the help of her teacher Annie Sullivan, and then spent much of her life lecturing to raise money for the teaching of the blind and deaf. Another example is Betty Ford, who was helped in her struggle with alcoholism and founded the Betty Ford Center to help others with addiction problems.

How can you make a difference? Many schools and companies are realizing the importance of community involvement and have appointed committees to find and organize volunteering opportunities. Make some kind of volunteering activity a priority on your schedule. Join a group from your company that tutors at a school. Organize a group of students to clean, repair, or entertain at a nursing home or shelter. Look for what's available to you or create opportunities on your own. Table 12-2 lists organizations that provide volunteering opportunities; you might also look into more local efforts or private clearinghouses that set up a number of different smaller projects.

Volunteerism is also getting a great deal of attention on the national level. The government has made an effort to stress the importance of community service as part of what it means to be a good citizen, and it provides support for that effort through AmeriCorps. AmeriCorps provides financial awards for education in return for community service work. If you work for AmeriCorps, you can use the funds you receive to pay current tuition expenses or repay student loans. You may work either before, during, or after your college education. You can find more information on AmeriCorps by contacting this organization:

The Corporation for National and Community Service
1201 New York Avenue, NW
Washington, D. C. 20525
1-800-942-2677

Sometimes it's hard to find time to volunteer when so many responsibilities compete for your attention. One solution is to combine other activities with volunteer work. Get exercise while cleaning a park or bring the whole family to sing at a nursing home on a weekend afternoon. Whatever you do, your actions will have a ripple effect, creating a positive impact for those you help and those they encounter in turn. The strength often found in people surviving difficult circumstances can strengthen you as well.

Valuing Your Environment

Your environment is your home. When you value it, you help to maintain a clean, safe, and healthy place to live. What you do every day has an impact on others around you and on the future. One famous slogan says that if you are not part of the solution, you are part of the problem. Every saved bottle, environmentally aware child, and reused bag is part of the solution. Take responsibility for what you can control—your own habits—and develop sound practices that contribute to the health of the environment.

		Table 12-2
AIDS-related organizations	Kiwanis/Knights of Columbus/ Lions Club/Rotary	Organizations that can use your help.
American Red Cross		
Amnesty International	Libraries	
Audubon Society	Meals on Wheels	
Battered women shelters	Nursing homes	
Big Brothers and Big Sisters	Planned Parenthood	
Churches, synagogues, temples, and affiliated organizations such as the YMCA/YWCA or YMHA/YWHA	Schools	
	Scouting organizations	
Educational support organizations	Share Our Strength/other food donation organizations	
Environmental awareness/support organizations such as Greenpeace	Shelters and organizations supporting the homeless	
Hospitals		
Hot lines	Sierra Club/World Wildlife Fund	

Recycle anything that you can. What can be recycled varies with the system set up in your area. You may be able to recycle any combination of plastics, aluminum, glass, newspapers, and magazines. Products that make use of recycled materials are often more expensive, but if they are within your price range, try to reward the company's dedication by purchasing the products.

Trade and reuse items. When your children have outgrown their crib, baby clothes, and toys, give away whatever is still usable. Give clothing you don't wear to others who can use it. Organizations like the Salvation Army may pick up used items in your neighborhood on certain days or when you make arrangements with them. Wrap presents in newspapers and decorate with markers. Use your imagination—there are many, many items that you can reuse all around you.

Respect the outdoors. Participate in maintaining a healthy environment. Use products that reduce chemical waste. Pick up after yourself. Through volunteering, voicing your opinion, or making monetary donations, support the maintenance of parks and the preservation of natural undeveloped land. Be creative. One young woman planned a cleanup of a local lakeside area as the main group activity for the guests at her birthday party (she joined them, of course). Everyone benefits when each person takes responsibility for maintaining the fragile earth.

Remember that valuing yourself is the base for valuing all other things. Improving the earth is difficult unless you value yourself and think you deserve the best living environment possible. Valuing yourself will also help you understand why you deserve to enjoy the benefits of learning throughout your life.

WHY IS COLLEGE JUST THE BEGINNING OF LIFELONG LEARNING?

Although it may sometimes feel more like a burden, being a student is a golden opportunity. As a student, you are able to focus on learning for a period of time, and your school focuses on you in return, helping you gain access to knowledge, resources, and experiences. Take advantage of the academic atmosphere by developing a habit of seeking out new learning opportunities. That habit will encourage you to continue your learning long after you have graduated, even in the face of the pressures of everyday life.

Learning brings change, and change causes growth. As you change and the world changes, new knowledge and ideas continually emerge. Absorb them so that you can propel yourself into the future. Visualize yourself as a student of life who learns something new every single day.

Here are some lifelong learning strategies that can encourage you to continually ask questions and explore new ideas.

Investigate new interests. When information and events catch your attention, take your interest one step further and find out more. If you are fascinated by politics on television, find out if your school has political clubs that you can explore. If a friend of yours starts to take yoga, try out a class with him. If you really like one portion of a particular class, see if there are other classes that focus on that specific topic. Turn the regretful, "I wish I had tried that," into the purposeful, "I'm going to do it."

Read books, newspapers, magazines, and other writings. Reading opens a world of new perspectives. Check out what's on the bestseller list at your bookstore. Ask your friends about books that have changed their lives. Stay on top of current change in your community, your state, your country, and the world by reading newspapers and magazines. A newspaper that has a broad scope, such as *The New York Times* or *Washington Post,* can be an education in itself. Explore religious literature, family letters, and Internet news groups and Web pages. Keep something with you to read for those moments when you have nothing to do.

Spend time with interesting people. When you meet someone new who inspires you and makes you think, keep in touch. Have a potluck dinner party and invite one person or couple from each corner of your life—your family, your work, your school, a club to which you belong, your neighborhood. Sometimes, meet for reasons beyond just being social. Start a book club, a home-repair group, a play-reading club, a hiking group, or an investing group. Get to know people of different cultures and perspectives. Learn something new from each other.

Pursue improvement in your studies and in your career. When at school, take classes outside of your major if you have time. After graduation, continue your education both in your field and in the realm of general knowledge. Stay on top of ideas, developments, structural changes, and new technology in your field by seeking out continuing education courses. Sign

TERMS

Continuing education
Courses that students can take without having to be part of a degree program.

up for career-related seminars. Take single courses at a local college or community learning center. Some companies offer additional on-the-job training or will pay for their employees to take courses that will improve their knowledge and skills. If your company doesn't, you may want to set a small part of your income aside as a "learning budget." When you apply for jobs, you may want to ask about what kind of training or education the company offers or supports.

Nurture a spiritual life. You can find spirituality in many places. You don't have to regularly attend a house of worship to be spiritual, although that may be an important part of your spiritual life. "A spiritual life of some kind is absolutely necessary for psychological 'health,'" says psychologist and author Thomas Moore in his book *The Care of the Soul*. "We live in a time of deep division, in which mind is separated from body and spirituality is at odds with materialism."[9] The words *soul* and *spirituality* hold different meaning for each individual. Decide what they mean to you. Whether you discover them in music, organized religion, friendship, nature, cooking, sports, or anything else, making them a priority in your life will help you find a greater sense of balance and meaning.

Experience what others create. Art is "an adventure of the mind" (Eugene Ionesco, playwright); "a means of knowing the world" (Angela Carter, author); something that "does not reproduce the visible; rather, it makes visible" (Paul Klee, painter); "a lie that makes us realize truth" (Pablo Picasso, painter); a revealer of "our most secret self" (Jean-Luc Godard, filmmaker). Through art you can discover new ideas and shed new light on old ones. Explore all kinds of art and focus on any forms that hold your interest. Seek out whatever moves you—music, visual arts, theater, photography, dance, domestic arts, performance art, film and television, poetry, prose, and more.

Make your own creations. Bring out the creative artist in you. Take a class in drawing, in pottery, or in quilting. Learn to play an instrument that you have always wanted to master. Write poems for your favorite people or stories to read to your kids. Invent a recipe. Design and build a set of shelves for your home. Create a memoir of your life. You are a creative being. Express yourself, and learn more about yourself, through art.

Lifelong learning is the master key that unlocks every door you will encounter on your journey. If you keep it firmly in your hand, you will discover worlds of knowledge—and a place for yourself within them.

HOW CAN YOU LIVE YOUR MISSION?

As you learn and change, so may your life's mission. Whatever changes occur, your continued learning will give you a greater sense of security in your choices. Recall your mission statement from Chapter 4. Think about how it is changing as you learn and develop. It will continue to reflect your goals, values, and strengths if you live with integrity, roll with the changes that come your way, continue to observe the role models in your life, and work to achieve your personal best in all that you do.

Live With Integrity

You've spent a lot of time exploring who you are, how you learn, and what you value. Integrity is about being true to that picture you have drawn of yourself while also considering the needs of others. Living with integrity will bring you great personal and professional rewards.

Honesty and sincerity are at the heart of integrity. Many of the decisions you make and act upon in your life are based on your underlying sense of what is "the right thing to do." Having integrity puts that sense into day-to-day action.

> **TERMS**
>
> **Integrity**
> Adherence to a code of moral values; incorruptibility, honesty.

The Marks of Integrity

A person of integrity lives by the following principles:

1. *Honest representation of yourself and your thoughts.* For example, you tell your partner when you are hurt over something that he or she did or didn't do.
2. *Sincerity in word and action.* You do what you say you will do. For example, you tell a co-worker that you will finish a project when she has to leave early, and you follow through by completing the work.
3. *Consideration of the needs of others.* When making decisions, you take both your needs and the needs of others into account. You also avoid hurting others for the sake of your personal goals. For example, your sister cares for your elderly father in her home where he lives with her. You spend three nights a week with him so that she can take a course toward her degree.

The Benefits of Integrity

When you act with integrity, you earn trust and respect from yourself and from others. If people can trust you to be honest, to be sincere in what you say and do, and to consider the needs of others, they will be more likely to encourage you, support your goals, and reward your hard work. Integrity is a must for workplace success. To earn promotions, it helps to show that you have integrity in a variety of situations.

> "And life is what we make it, always has been, always will be."
>
> GRANDMA MOSES

Think of situations in which a decision made with integrity has had a positive effect. Have you ever confessed to an instructor that your paper is late without a good excuse, only to find that despite your mistake you have earned the instructor's respect? Have extra efforts in the workplace ever helped you gain a promotion or a raise? Have your kindnesses toward a friend or spouse moved the relationship to a deeper level? When you decide to act with integrity, you can improve your life and the lives of others.

Most importantly, living with integrity helps you believe in yourself and in your ability to make good choices. A person of integrity isn't a perfect person but one who makes the effort to live according to values and principles, continually striving to learn from mistakes and to improve. Take responsibility for making the right moves, and you will follow your mission with strength and conviction.

Roll With the Changes

Think again about yourself. How has your idea of where you want to be changed since you first opened this book? How has your self-image changed? What have you learned about your values, your goals, and your styles of communication and learning? Consider how your educational, professional, and personal goals have changed. As you continue to grow and develop, keep adjusting your goals to your changes and discoveries.

Stephen Covey says in *The Seven Habits of Highly Effective People*, "Change—real change—comes from the inside out. It doesn't come from hacking at the leaves of attitude and behavior with quick fix personality ethic techniques. It comes from striking at the root—the fabric of our thought, the fundamental essential paradigms which give definition to our character and create the lens through which we see the world."[10]

Examining yourself deeply in that way is a real risk. Most of all, it demands courage and strength of will. Questioning your established beliefs and facing the unknown are much more difficult than staying with how things are. When you have the courage to face the consequences of trying something unfamiliar, admitting failure, or challenging what you thought you knew, you open yourself to growth and learning opportunities. You can make your way through changes you never anticipated if you make the effort to live your mission—in whatever forms it takes as it changes—each day, each week, each month, and for years to come.

TERMS
Paradigm
An especially clear pattern or typical example.

Learn From Role Models

People often derive the highest level of motivation and inspiration from learning how others have struggled through the ups and downs of life and achieved their goals. Somehow, seeing how someone else went through difficult situations can give you hope for your own struggles. The positive effects of being true to one's self become more real when an actual person has earned them.

Learning about the lives of people who have achieved their own version of success can teach you what you can do in your own life. Bessie and Sadie Delany, sisters and accomplished African-American women born in the late 1800s, are two valuable role models. They took risks, becoming professionals in dentistry and teaching at a time when women and minorities were often denied both respect and opportunity. They worked hard to fight racial division and prejudice and taught others what they learned. They believed in their intelligence, beauty, and ability to give, and lived without regrets. Says Sadie in their *Book of Everyday Wisdom*, "If there's anything I've learned in all these years, it's that life is too good to waste a day. It's up to you to make it sweet."[11]

TERMS
Role model
A person whose behavior in a particular role is imitated by others.

Aim for Your Personal Best

Your personal best is simply the best that you can do in any situation. It may not be the best you have ever done. It may include mistakes, for nothing significant is ever accomplished without making mistakes and taking risks. It may shift from situation to situation. As long as you aim to do your best, though, you are inviting growth and success.

Aim for your personal best in everything you do. As a lifelong learner, you will always have a new direction in which to grow and a new challenge to face. Seek constant improvement in your personal, educational, and professional life, knowing that you are capable of that improvement. Enjoy the richness of life by living each day to the fullest, developing your talents and potential into the achievement of your most valued goals.

Kaizen is the Japanese word for "continual improvement." Striving for excellence, always finding ways to improve on what already exists, and believing that you can impact change are at the heart of the industrious Japanese spirit. The drive to improve who you are and what you do will help to provide the foundation of a successful future.

Think of this concept as you reflect on yourself, your goals, your lifelong education, your career, and your personal pursuits. Create excellence and quality by continually asking yourself, "How can I improve?" Living by *kaizen* will help you to be a respected friend and family member, a productive and valued employee, and a truly contributing member of society. You can change the world.

Chapter 12 Applications

Name _____ Date _____

KEY INTO YOUR LIFE
Opportunities to Apply What You Learn

 Questions in Science

Read the following article by Kathy Svitil from *Discover* magazine's Web site[12] (www.discover.com), and then answer the questions that follow.

Brave New Genes: Are We Ready for a World of Custom Chromosomes and Designer DNA?

For about four billion years, life has progressed through a series of changes mediated by intrinsic genetic variation and natural selection. Genetic engineering offers a radical new possibility: now a species can tinker with its own DNA. "We're seizing control of our evolution, in some sense," said Gregory Stock, the director of the Program on Science, Technology and Society at the University of California at Los Angeles. Stock and John Campbell, an evolutionary biologist at the UCLA School of Medicine, organized "Engineering the Human Germline" to grapple with the implications of that power. The March 20 symposium was the first forum where scientists and ethicists gathered before the public to discuss the biological, technological, and ethical issues involved in redesigning the human genetic code.

The UCLA symposium marked a new stage in the debates described in the May issue of *Discover*. The focus was on germ-line gene therapy, which involves manipulation of DNA in sperm and eggs (germ cells) and produces permanent, inheritable genetic alterations. This contrasts with somatic gene therapy, a more limited technique in which genes are introduced just into particular tissue cells or cell lines that will not be passed along to off-spring. Unlike somatic therapy, germ-line therapy is not yet a scientific reality. But "the technology is moving so quickly that something is likely to be possible in the next few decades," said Stock. "We need to discuss it now, so that it doesn't spring unexpectedly upon us—and the discussion shouldn't go on tucked into the recesses of a university."

Talk of genetic engineering inevitably inspires Brave New World scenarios: If the technology exists to wipe disease genes forever from a family tree, what is to stop parents from making a few extra changes to extend their child's life span, or to alter personality, or to increase height, strength, and attractiveness?

As the participants at UCLA pointed out, those questions are not so far-fetched. Germ-line therapy is already old hat in animals. Back in the early 1980s, conference participant Leroy Hood, now of the University of Washington, along with his California Institute of Technology colleagues, added new genes to mice embryos who had a deadly hereditary nervous system disorder. The gene treatment cured the mice, along with their subsequent offspring.

Researchers are rapidly developing new genetic technologies for use in humans as well. Several researchers recently announced the creation of artificial chromosomes—structures that contain no genetic material but that have docking sites onto which genes could be added. Artificial chromosomes could be slipped into cells, where, like normal chromosomes, they would replicate and pass their attached genetic messages from parent cell to daughter cell.

Campbell envisions using artificial chromosomes to confer protection against diseases like AIDS or specific forms of cancer. For instance, Campbell outlined a scheme by which prostate cancer could be targeted using special "cassettes" of genes that instruct the body to produce a toxin that attacks the cells in the prostate. Other genes would ensure that the toxin gene is activated, or expressed, only in the presence of an insect hormone, ecdysone. These genes would be placed on artificial chromosomes and inserted into early embryonic cells. If a man were later diagnosed with prostate cancer, he could be given a shot of ecdysone, triggering the toxin. The body would then kill off the prostate cells—and the cancer along with them. Other artificial chromosomes could be fabricated to deal with different health threats, or even to make desirable but nonessential genetic changes.

That's the theory, at least. Will artificial chromosomes work in real trials? Geneticist Mario Capecchi of the University of Utah sounded optimistic. On purely technical grounds, he noted, "germ-line therapy is actually much simpler than somatic therapy." But that's not saying much. Somatic gene therapy was first tested on humans back in 1990, when W. French Anderson of the University of Southern California used it to treat a young girl with an inherited immune system disorder. Since then, thousands of patients with a myriad of diseases have received somatic therapy. It has yet to cure even one.

Despite that dismal record, Anderson argued that long-term experience with somatic therapy is a necessary precondition before scientists should attempt germ-line therapy. He also wants to make sure that germ-line techniques prove safe and reliable in lab monkeys, and that the public understands and approves of the procedure. "In principle, I'm for it on the most fundamental of grounds: human nature. None of us wants to pass on to our children lethal genes if we can prevent it. And that will drive germ-line therapy. It is going to happen. The issue is when is it safe, and when is it ethical."

Anderson is particularly concerned about the proposed use of germ-line therapy for "enhancement" rather than for curing disease. At the UCLA symposium, scientists discussed the possibility of adding genes to increase height, change hair color, or make a person more emotionally stable. During some of those talks, "I had a hard time sitting in my chair," Anderson said. "There has been so much talk about 'improving' and 'enhancing,' where enhancement means going from normal to above normal—and I don't know what normal is." That sentiment was echoed by bioethicist John Fletcher of the University of Virginia, who collaborated with Anderson on a seminal 1980 paper on the ethics of gene therapy in humans.

"There is a distinction between being a fool for genetic engineering and being a damn fool for genetic engineering," Fletcher said. "I would like to see investigators turn their attention to therapy."

Calls for a cautious approach to gene therapy, guided by government regulation, met with scorn from James Watson, president of the Cold Spring Harbor Laboratory on Long Island, New York. Watson won a Nobel prize in 1962 for his work deciphering the structure of DNA and establishing the Human Genome Project. "If we wait for the success of somatic before trying germ-line, we risk the sun burning out," said Watson, who was equally critical of blanket statements that all genetic enhancement is a bad idea. "If we can make better human beings by knowing how to add genes, why shouldn't we do it?" he asked. "The biggest ethical problem we have is not using our knowledge."

Source: Kathy A. Svitil, © 1998. Reprinted with permission of *Discover Magazine*.

1. What ethical questions are raised? _____

_____

_____

2. What legal questions are raised? _____
_____
_____
_____
_____

3. What economic questions are raised? _____
_____
_____
_____
_____

4. What is your personal reaction? _____
_____
_____
_____
_____

5. As a person of science, what is your reaction? Is it different from your personal reaction and, if so, in what ways? _____
_____
_____
_____
_____

12.2 Looking at Change, Failure, and Success

Life can go by so fast that you don't take time to evaluate what changes have taken place, what failures you could learn from, and what successes you have experienced. Take a moment now and answer the following questions for yourself.

What are the three biggest changes that have occurred in your life this year?

1. *New Car*
2. *Got raise on job*
3. *Learning to play instrument - piano.*

Choose one that you feel you handled well. What shifts in priorities or goals did you make?

New car - because it has brought more responsibilityes to my life.

Choose one that you could have handled better. What happened? What do you think you should have done?

Playing the piano - even though I've learn quite alot, I could have spend more time learning instead of going out.

Now name a personal experience, occurring this year, that you would consider a failure. What happened?

not being able to quit smoking

How did you handle it—did you ignore it, blame it on someone else, or admit and explore it?

well I could not resist the urge.

What did you learn from experiencing this failure?

~~I will definately get lung cancer~~ Chances of getting lung can have increased

Finally, describe a recent success of which you are the most proud.

I got a B on my career research paper.

How did this success give you confidence in other areas of your life?

I feel a little more confident about my writing skills are not

12.3 Volunteering

Research volunteering opportunities in your community. What are the organizations? What are their needs? Do any volunteer positions require an application, letters of reference, or background checks? List three possibilities for which you have an interest or a passion.

1. _____
2. _____
3. _____

Of these three, choose one that you feel you will have the time and ability to try next semester. Suggestions that don't take up too much time include spending an evening serving in a soup kitchen or driving for Meals on Wheels during a lunch or dinner shift. Name your choice here and tell why you selected it.

_____
_____
_____
_____

Research the suggestion you have chosen. Describe the activity. What is the time commitment? Is there any special training involved? Are there any problematic or difficult elements to this experience?

_____
_____
_____
_____
_____
_____
_____

12.4 Lifelong Learning

Review the strategies for lifelong learning in this chapter. Which ones mean something to you? Which do you think you can do, or plan to do, in your life now and when you are out of school? Name them and briefly discuss the role they play in your life.

_____
_____
_____
_____
_____

KEY TO SELF-EXPRESSION
Discovery Through Journal Writing

To record your thoughts, use a separate journal or the lined page at the end of the chapter.

Ethical Questions

Take some time to write about questions you have concerning science and technologies discovery and use.

Moving Ahead CHAPTER 12

Name _____ Date _____

Journal

Journal

Name _____ Date _____

Endnotes

Chapter 1

[1] U. S. Department of Education, National Center for Education Statistics (NCES), *1996 Digest of Education Statistics*, Table 266 and Table 267.

[2] Ibid., Table 266.

[3] Ibid., Table 266.

[4] Ibid., Table 267.

[5] U. S. Department of Education, National Center for Education Statistics, *Nontraditional Undergraduates: Trends in Enrollment from 1986 to 1992 and Persistence and Attainment Among 1989-90 Beginning Postsecondary Students*, NCES 97-578, by Laura J. Horn. Project Officer Dennis Carroll, (Washington, D. C.: U. S. Government Printing Office, 1996), 26.

[6] NCES *1996 Digest of Education Statistics*, Table 177.

[7] U. S. Department of Education, National Center for Education Statistics, *The Condition of Education 1996*, NCES 96-304, by Thomas M. Smith (Washington, D. C.: U. S. Government Printing Office, 1996), 60–61.

[8] H. N. Fullerton, Jr., (1995). "The 2005 Labor Force: Growing, But Slowly," *Monthly Labor Review 118*(11): 29–44.

[9] *1998-1999 Occupational Outlook Handbook*, Bureau of Labor Statistics, U. S. Department of Labor.

[10] National Science Foundation, Division of Human Resources, Directorate for Education and Human Resources, National Science Foundation, www.nsf/Women and Girls.org.

[11] Richard W. Riley, Department of Education, *Remarks on TIMSS Results: Impact for Our Future and Individual Opportunities* [online]. Available: http://www.ed.gov/press-releaes/02-1998/timss.html, February 1998.

[12] Ibid.

[13] *Occupational Outlook Handbook 1998-99*.

[14] Ibid.

[15] *Washington State University 1996-1998 Catalog*. (Washington State University, Pullman Washington, 1996), 29.

Chapter 2

[1] David Sobel, "Among Planets: In the Age of the Hubble Telescope and Rocks From Mars, Planetary Exploration Is Getting New Respect," *The New Yorker* (Dec. 9, 1996): 84–90.

[2] J. R. Katz. *Majoring in nursing: From prerequisites to post graduate and beyond.* (New York: Farrar, Strauss & Giroux, 1999.)

[3] Ibid., 90.

[4] *1998–1999 Occupational Outlook Handbook*, Bureau of Labor Statistics.

[5] N. A. Campbell, *Biology*, 4th ed. (Menlo Park, CA: Benjamin/Cummings, 1997), 22–24.

[6] *1998–1999 Occupational Outlook Handbook*.

[7] Ibid.

[8] J. Garcia, *Majoring in Engineering: How To Get From Your Freshman Year to Your First Job* (New York: Noonday Press, 1995), 32.

[9] Ibid., 36.

[10] Ibid., 38.

[11] *1998–1999 Occupational Outlook Handbook*.

[12] "Program for Women and Girls in Science, Engineering, and Mathematics," National Science Foundation [online]. Available: http://www.her.nsf.gov/HER/HRD/women/guidelines.htm, 1998.

Chapter 3

[1] *The Third International Mathematics and Science Study (TIMSS)*, 1998, National Center for Education Statistics.

[2] Richard W. Riley, Remarks to the Conference of American Mathematical Society and Mathematical Association of America, "The State of Mathematics Education: A Strong Foundation for the 21st Century," January 8, 1998.

[3] Barbara Soloman, North Carolina State University, Raleigh, NC.

[4] Howard Gardner, *Multiple Intelligences: The Theory in Practice* (New York: HarperCollins, 1993), 5–49.

[5] Joyce Bishop, Ph.D., Psychology faculty, Golden West College, Huntington Beach, CA.

Chapter 4

[1] Paul R. Timm, Ph.D., *Successful Self-Management: A Psychologically Sound Approach to Personal Effectiveness* (Los Altos, CA.: Crisp Publications, Inc., 1987), 22–41.

[2] Stephen Covey, *The Seven Habits of Highly Effective People* (New York: Simon & Schuster, 1989), 70–144, 309–318.

Chapter 5

[1] Frank T. Lyman, Jr., Ph.D., "Think-Pair-Share, Thinktrix, Thinklinks, and Weird Facts: An Interactive System for Cooperative Thinking." In *Enhancing Thinking Through Cooperative Learning*, ed. Neil Davidson and Toni Worsham (New York: Teachers College Press, 1992), 169–181.

[2] Roger von Oech, *A Kick in the Seat of the Pants* (New York: Harper & Row Publishers, 1986), 5–21.

[3] R. M. Roberts, *Serendipity: Accidental Discoveries in Science* (New York: John Wiley & Sons, 1989, ix.

[4]Ibid., 164.
[5]Ibid., x.
[6]"1998 Discover Technology Awards." *Discover* 19 (7): 46.
[7]Ibid., 51.
[8]R. M. Roberts, *Serendipity*, x.
[9]Dennis Coon, *Introduction to Psychology: Exploration and Application*, 6th ed. (St. Paul: West Publishing Company, 1992), 295.

Chapter 6

[1]U. S. Department of Education, National Center for Education Statistics, *The Condition of Education 1996*, NCES 96–304, by Thomas M. Smith (Washington, D. C.: U. S. Government Printing Office, 1996), 84.

[2]Sherwood Harris, *The New York Public Library Book of How and Where to Look It Up* (Englewood Cliffs, NJ: Prentice Hall, 1991), 13.

[3]George M. Usova, *Efficient Study Strategies: Skills for Successful Learning* (Pacific Grove, CA: Brooks/Cole Publishing Company, 1989), 45.

[4]Francis P. Robinson, *Effective Behavior* (New York: Harper & Row, 1941).

[5]Sylvan Barnet and Hugo Bedau, *Critical Thinking, Reading, and Writing: A Brief Guide to Argument*, 2nd ed. (Boston: Bedford Books of St. Martin's Press, 1996), 15–21.

[6]P. S. Fardy and F. G. Yanowitz, *Cardiac Rehabilitation, Adult Fitness, and Exercise Testing*, 3rd ed. (Baltimore: Williams & Wilkins, 1995), 246–247.

[7]John J. Macionis, *Sociology*, 6th ed. (Upper Saddle River, NJ: Prentice Hall, 1997), 174.

[8]Teresa Audesirk and Gerald Audesirk, *Life on Earth* (Upper Saddle River, NJ: Prentice Hall, 1997), 55–56.

Chapter 7

[1]Walter Pauk, *How to Study in College*, 5th ed. (Boston: Houghton Mifflin Company, 1993), 110–114.

[2]Analysis based on Lynn Quitman Troyka, *Simon & Schuster Handbook for Writers* (Upper Saddle River, NJ: Prentice Hall, 1996), 22–23.

Chapter 8

[1]Ralph G. Nichols, "Do We Know How to Listen? Practical Helps in a Modern Age," *Speech Teacher* (March 1961): 118–124.

[2]Ibid.

[3]Herman Ebbinghaus, *Memory: A Contribution to Experimental Psychology*, trans. H. A. Ruger and C. E. Bussenius (New York: New York Teacher's College, Columbia University, 1885).

[4]Many of the examples of objective questions used in this chapter are from Gary W. Piggrem, Test Item File for Charles G. Morris, *Understanding Psychology*, 3rd ed. (Upper Saddle River, NJ: Prentice Hall, 1996).

Chapter 9

[1]Susan Milius, "Why Florida's Cormorants Look Drunk," *Science News* Vol. 154, 1998.

Chapter 10

[1] Edith Wharton, "False Dawn," in *Old New York* (New York: Simon & Schuster, 1951), 18–19.

[2] Sheryl McCarthy, *Why Are the Heroes Always White?* (Kansas City: Andrews and McMeel, 1995), 188.

[3] John Hockenberry, *Moving Violations* (New York: Hyperion, 1995), 78.

[4] Tamera Trotter and Joycelyn Allen, *Talking Justice: 602 Ways to Build and Promote Racial Harmony* (Saratoga, CA: R & E Publishers, 1993), 51.

[5] Sheryl McCarthy, *Why Are the Heroes?* . . . , 137.

[6] Louis E. Boone, David L. Kurtz, and Judy R. Block, *Contemporary Business Communication* (Englewood Cliffs, NJ: Prentice Hall, 1994), 49–54.

[7] Adapted by Richard Bucher, Professor of Sociology, Baltimore City Community College, from Paula Rothenberg, William Paterson College of New Jersey.

Chapter 11

[1] U. S. Department of Education, National Center for Education Statistics, *Profiles of Undergraduates Who Work: 1995–96*, NCES 98-084, by Laura J. Horn, Mark D. Premo, Andrew G. Malizio, Project Officer, and MPR Associates, Inc. (Washington, D. C.: U. S. Government Printing Office, May 1998), 15.

[2] U. S. Department of Education, National Center for Education Statistics, *Digest of Education Statistics 1996*, NCES 96-133, by Thomas D. Snyder. Production Manager, Charlene M. Hoffman. Program Analyst, Claire M. Geddes (Washington, D. C: U. S. Government Printing Office, 1996), 320–321.

[3] Ibid., 324–325.

[4] U. S. Department of Education, *The 1998–99 Student Guide to Financial Aid*.

Chapter 12

[1] L. A. Whittaker, "The Implications of the Human Genome Project for Family Practice," *Journal of Family Practice* 35 (3) (1992): 294–301.

[2] D. C. Wertz, "Ethical and Legal Implications of the New Genetics: Issues for Discussion. *Social Science and Medicine* (1992): 495–505.

[3] Ibid.

[4] March of Dimes, *Genetic Testing and Gene Therapy* (New York: March of Dimes, 1992).

[5] D. C. Wertz, Ethical and Legal Implications.

[6] T. H. Maugh, "When It Comes to Romance, Forget DNA," *The Los Angeles Times*, in *The Spokesman Review*, pp. 1, 8, December 3, 1994.

[7] B. P. Sachs and B. Korf, "The Human Genome Project: Implications for the Practicing Obstetrician." *Obstetrics and Gynecology* (1993): 458–462.

[8] Isaac Asimov, "My Own View," in *The Encyclopedia of Science Fiction*, ed. Robert Holdstock (1978).

[9] Thomas Moore, *The Care of the Soul* (New York: Harper Perennial, 1992), xi–xx.

[10] Stephen Covey, *The Seven Habits of Highly Effective People* (New York: Simon & Schuster, 1989), 70–144, 309–318.

[11] Sarah Delany and Elizabeth Delany with Amy Hill Hearth, *Book of Everyday Wisdom* (New York: Kodansha International, 1994), 123.

[12] Kathy Svitil, "Brave new genes: Are we ready for a world of custom chromosomes and designer DNA?" *Discover* Web page, www.discover.com, May 20, 1998.

Index

abbreviations, 174
abilities, 72–73
absolute truth, 296
abstract, 187
   thinking, 42
academic:
   advisors, 10
   assistance, 10–11
   directory, 14
   writing, 139
accepting and dealing with differences, 246
accomplishments, 73, 89
achieving goals, 89
acronyms, 204
action, 140
   verbs, 214
active:
   learners, 54, 63–64
   listener, 198
   readers, 153
   /reflective learners, 54
activities, 91
addressing stereotypes, 242
ADHD, 5, 141, 180, 200
administrative personnel, 12
adult education services, 13
adults returning to school, 13
adventurers, 63
advisors and counselors, 12
African-American students, 4
aggressive communicators, 252, 263
agricultural scientist, 32
Alaskan native students, 4
alertness to real world, 48
almanacs, 156

alternative careers in science, 42
American Indian students, 4
American Institute of Biological Sciences, 32
AmeriCorps, 306
analogy, likeness, comparison, 113–114
analysis, 116
   of arguments, 114
APA Publication Manual, 184
apply what you learn, 100, 132, 158, 191, 217, 233, 261, 290, 313
appreciation, 254
   of differences, 248
   of yourself, 304
aquatic biologist, 31
Asian students, 4
assertive:
   behavior, 252
   communicators, 263
assess and analyze, 118
assessment, 51–60, 249
associations, 204
astronomist, 37
attendance, 227
Attention deficit hyperactivity disorder (ADHD), 5, 141, 180, 200
audience, 176–177, 181, 186, 250
audiologist, 36

bad habits, 71
barriers:
   to communication, 240
   to understanding, 240
   women face in the technical sciences, 42
basic skills, 271
beachcombers, 130
biases, 135
bibliographies, 156
biochemists, 31
biographical reference works, 156
biological science, 29–30, 32
biology, 30–32
   careers in, 31
biomedical equipment technician, 33
biotechnology, 231
Bishop, Joyce, 158, 160
blaming:
   others, 301
   yourself, 302
bodily-kinesthetic:
   intelligence, 58
   learners, 63
body of a paper, 183, 186
book catalog, 156
botanists, 31
brain research, 112
brain-scanning technology, 113
brainstorming, 118, 130, 172, 178–179, 256
   in a group, 133
   on the idea wheel, 132
budgeting, 283–286
bulletin boards, 11

calendar, 96–97
career:
   analysis, 21
   in biology, 32

in computer science, 40
exploration, 267
field, 269
goals, 270
in the health sciences, 33–35
managing, 267–271
in mathematics, 39
options, 267
path, 268
in physical science, 37–38
placement registry, 276
planning and placement office, 11, 274
CareerPath.com, 276
careers:
in biology, 30
in computer science, 40
in engineering science, 41
in the health sciences, 33–35
in physical science, 37–38
categories, 29
categorization, 115
cause and effect, 114, 154, 211, 302
analysis, 116
causes of problem, 121
CD-ROM databases, 157
chai, 99
challenges on the job, 275
change, 296–299, 311
brings different needs, 298
failure, and success, 315
of majors, 45
produces new priorities, 299
chapter headings, 160
checklist, 181, 187
chemical engineer, 41
chemist, 38
child-care center, 13
choices, 7
choosing and evaluating values, 82
civil engineer, 41
clarity and conciseness, 186
class notes, 168
classification, 115
sciences, 29
classified ads, 276
closed-minded approach vs. open-minded approach, 241
college catalog, 14, 16
commencement, 293
commitment, 84, 272
to your audience, 177
common ground, 247

communication, 17, 124
aggressive, 252, 263
in the lab, 232
problems, 251
skills, 230, 271
style, 248–249, 250, 252, 262
success strategies, 252
through visual images, 204
community, 305
involvement, 9, 306
comparison, 113–114
comprehension, 142–143, 147, 163
components of, 145
methods for increasing, 142
and speed, 142
computer:
science, 30, 39
scientist, 40
conceptual or factual errors, 216
conceptualization, 115
conclusion, 183–184, 186
confidentiality, 295
conflict, 255–256
strategies, 255
consequences, 114
constructive criticism, 256–257
contacts, 274
content, 143
continuing education, 308
contrasting, 114
Cornell note-taking system, 172–173
counselors, 12
course:
catalog, 15
syllabus, 226
Covey, Stephen, 84, 311
creative:
expression, 131
innovations, 128
people in science, 128
thinkers, 131
thinking, 107, 127
creativity, 27, 111, 127, 129–131, 225, 272
in science, 127–128
credit card, 287–291
debt, 288
offers, 289
credit:
history, 287
hours, 226
crediting authors and sources, 184

critical:
and creative thinking, 107
questions, 112
reading, 137, 139, 145, 151, 155
critical thinking, 107–119, 126, 131, 153, 185–186, 207, 243, 282
advantages of, 111
to avoid errors, 210
material, 242
mind actions, 148
processes, 117
questions for career investigation, 269
response, 110
in science, 111
is a skill, 108
skills, 202, 231, 256
strategies, 210
critical writer, 184–185
criticism, 255–258
constructive, 256–257
cue column, 172
cultural:
backgrounds, 17
and communication differences, 4
curiosity, 228

data, collecting, 1
date books, 92, 94, 96, 102
decision making, 113, 117, 119–123, 271
process, 119–123
deductive reasoning, 115
dental:
assistant, 33
hygienist, 33
laboratory technician, 33
dentist, 33
depression, 113
diagnostic ultrasound technician, 36
DIALOG Information System, 157
dictionaries, 156
dietetic technologist, 33
dietitian, 33
differences, 114, 154
accepting and dealing with, 246
fear of, 246
diligence, 27
dimensions of learning, 54

disability, 141, 239, 244
discipline-specific publications, 144
discoveries in science, 129
discovering science, 25
discrimination, 19, 240, 243–244
discussion, 190
distinction, 114
distraction, external, 140
distractions, 140, 198–199
diverse:
    student body, 3–4
    student population, 4
    working world, 125
    world, 16, 237, 248
diversity, 16–17, 19, 237–240, 245–247
    discovery of, 261
    positive effects of, 239–240
    and teamwork, 17
    in your world, 238
divided attention, 198
docendo discimus, 232
down time, 97
drafting, 182
dreams, 81
dyslexia, 5, 141

earning potential, 8
ecologist, 32
economic:
    implications, 295
    instability, 297
ecosystems, 30
editing, 187
education:
    and employment, 7
    and income, 6
    increases choices and power, 8
    in science, 1
educational:
    goals, 88
    requirements, 46
effects, 120
    of problem, 121
electrical engineer, 41
electrocardiograph (ECG) technician, 34
electroencephalogram technician (EEG), 34
electronic:
    databases, 157
    planner, 92
    research, 157

elements of effective writing, 176
emergency medical technician (EMT), 34
employability, 8
employment agencies, 276
empowerment, 247
encoding stage, 200
encyclopedias, 156
engineering science, 30, 41
enrollment information, 14
entertaining, 177
environment, 306–307
environmental:
    biologist, 31
    issues, 31
equipment, 224
    you need in the lab, 226
errors, conceptual or factual, 216
essay:
    questions, 213–214
    tests, 215
ethical:
    implications, 295
    questions, 318
ethics in science, 31
ethnic background(s), 261
ethnocentrism, 17
evaluate, 119–120
    your notes, 191
evaluation, 116, 121, 123, 145, 154
evidence, 183
exam, overview of, 209 (*see also* Test)
example to principle, 115, 154
exercise prescription, 158
experiments, 223
explore:
    possible solutions, 118
    potential careers, 269
    your values, 100
exploring:
    potential majors, 44
    your options, 25
express yourself effectively, 248
external:
    characteristics, 247
    distractions, 198

facts, sequence, and description, 113
factual learners, 54, 64
failure, 300–304
    and success, 301

family contribution, expected, 280
fear of differences, 246
federal grant programs, 280
feedback, 256
feeler, 249, 262
female science majors, 5
field study program, 124
fields of study, 25
    and careers in science, 29
file card, 277
financial:
    aid, 278–281
    aid office, 10
    assistance, 10–11
    management, 291
    philosophy, 282
    stability, 282
first draft, 182–184
flashcards, 203, 205
flexibility, 275, 293, 297–298
    vs. inflexibility, 298
forgiving yourself, 302
freewriting, 179, 182
    your draft, 182
FSEOG (Federal Supplemental Educational Opportunity Grants), 280
future, building, 293, 302

Gardner, Howard, 58
general principles reading, 145
generalization, 115
genetic engineering, 30
genetics, 31, 231, 294–295
    studies, 295
geologist, 38
giver, 63
global:
    communities, 296
    economy, 16
gloves, 226
goals, 82–95, 98, 111, 119–120, 299
    achieving, 84, 89
    for the future, 126
    linking, 87
    in science, 81
    in time, 84
    with values, 87
    within particular time frames, 85
goal setting, 81–88
golden rule, 254
good habits, 71

good writing, in science, 175
Gould, Stephen Jay, 26
graduation, 293
grant writing, 27
grants, 27, 280
    federal, 280
    and scholarship money, 5
group, 258–260, 264
    interaction, 258
    members, 259–260
    project, 259
guessing, 210, 212
guided:
    inquiry, 225
    notes, 171

habits, 71–72, 77–78
handbook, 14, 16
hands-on experience, 9
Hawking, Stephen W., 26
health:
    science, 32–37
    services administrator, 34
healthy living, 33
hearing loss, 200
highlighting, 149–151
Hispanic students, 4
historical documents, 139
holistic learners, 55, 66
Holocaust, 248
honesty and sincerity, 310
Hubble telescope, 29
human:
    interaction, 238
    nature, 241
    resources, 126
human genome, 31
    project, 294
hypotheses, 118, 224–225

"I'll work for free," 9
"I" messages, 251
implement, 118
imprint on the world, 305
improving recall, 201
income and employment, 7
inductive reasoning, 115
inflexibility, 298
information superhighway, 157
informative writing, 176
inquiry, 117, 226
    in action, 233
    -based lab science, 224
    -based research, 224

in science, 112
instructors, 9, 12
    cues, 169
    as an important resource, 151
    teaching styles of, 69
integrity, 272, 310
intelligence, 58
interaction with people, 237
interest in science, 29
interests, 70–71
    majors, and careers, 49
interlibrary loan, 156
internal distractions, 198
international testing scores, 52
Internet, 20, 157, 234, 276, 297
interpersonal:
    intelligence, 58
    learners, 63
    relationships, 60, 62
interview, 277–278
intrapersonal intelligence, 58
introduction, 183, 186–187
intuitor learning style, 248, 250, 262

jealousy, 243
job:
    /career assistance, 10–11
    -hunting strategies, 274
    information, 274
    listings, 276
    priorities, 290
    satisfaction, 267
    -specific skills, 8
joie de vivre, 46
journal writing exercise, 22, 49, 78, 104, 135, 164, 194, 220, 235, 264, 291, 318
journalists' questions, 180
journals, 144
judgment, 116

kaizen, 312
Keirsey and Bates, 60
kente, 260
key words, phrases, and concepts, 148
knowledge and skills in science, 5
krinein, 131

lab:
    books, 226
    classes, 223, 227
    coat, 226

    equipment, 226
    instructions, 232
    rules, 229
    safety, 228–229
    success, 231
    time, 226
    write-up, 228–229
laboratory, working in, 223–231 (*see also* Lab)
leader, 258, 260
leadership, 272
    strategies, 259
learn from your mistakes, 215
learning, 51
    disabilities, 5, 141, 180, 198, 200
    effectively, 54
    from others, 47
    from your mistakes, 219
    style profile, 68, 73, 78
learning styles, 53–68, 249
    inventory, 53–54, 57, 75
    profile, 62, 68, 73, 78
    techniques, 68
learning-disabled student, 240
legal implications, 294
levels of inquiry, 225
library, 155
    book catalog, 156
    computer records of, 156
    reference work, 12
    tours, orientation, training sessions, 155
licensed practical nurse (LPN), 34
life changes, 82
lifelong learning, 308–309, 318
lifestyles, 239
limitations, 72–73
linear learners, 55, 66
linking goals together, 87
linking majors to career areas, 45
listener's style, 249
listening, 197–199, 217
    challenges, 198
    conditions, 217
    environment, 198–199
    is helped by, 199
    is hindered by, 199
    research, 198
    skills, 8
    style, 249
    techniques, 200
literacy, 137

literature review, 190
loan, 278–279
    application, 279
logical-mathematical intelligence, 58
long-term:
    goals, 81, 85–87, 92–93
    memory, 201

machine-scored tests, 210
major, 44–46
manage your time, 89
managing distractions, 140
mapping your course, 81
marketable skills, 88
math skills, 232
mathematics, 27, 30, 39
MBTI, 249
McClintock, Barbara, 26
mechanical engineer, 41
medical:
    assistant, 34
    laboratory technician/medical technologist, 34
    social worker, 34
memorization, 201
memory, 197, 200–204
    improvement strategies, 201
    principles, 218
    skills, 197, 204
    techniques, 201, 203–204, 216
    tool, 203
mentors, 12, 290
meteorologist, 37
meteorology, 38
methods, 190
microbiology, 31
mind actions, 113, 116–117, 153–154, 186
mind map, 172
mission, 309
    statement, 84, 87, 309
mnemonic device, 116, 203–204, 218–219
modify, 300
money management, 267, 282–287
    budgeting, 283–285
    credit cards and, 287
    saving, 286
MRI, 36
multicultural communication, 272
multiculturalism, 17

multiple intelligences, 67
    theory, 58
multiple-choice questions, 211, 213
musical intelligence, 58
musical/rhythmic people, 67
Myers-Briggs, 60
    Type Indicator (MBTI), 249
    types, 54

narrating, 177
narrative style, 143
narrow your topic, 178
National Center for Education Statistics, 3
National Science Foundation, 5, 42
naturalistic:
    intelligence, 58
    learners, 67
    people, 67
negative effects, 120–121, 123
networking, 274
Nobel prize, 26
note-taking, 167–172
    skills, 167
    system, 170, 172, 191
notes, 173, 191
    during class discussions, 169
    during a lecture, 169
    in the margins, 148–149
    in outline form, 170
    as a valuable study tool, 170
nuclear medicine technician, 35
nurse practitioner, 34
nurses, 36
nursing assistant certified (NAC or CNA), 34

obesity, 244
objective questions, 211
observation, 28
    as a critical skill, 112
    skills, 28, 232
observe, 118
obstacles, 110
Occupational Outlook Handbook, 6
occupational therapist, 35
older students, 4
on-line:
    search capacity, 156
    services, 276
ophthalmic technologist, 35
opportunities to apply what you

learn, 75
optician, 35
organizations, 13–14, 307
    and clubs, 10
organizer, 63–64
orthotic:
    assistant, 35
    technician, 35
orthotist/prosthetist, 35
outlining, 170–171, 181
overview of the exam, 209

paraphrase, 184
participant, 258
participation strategies, 259
passive communicators, 252, 263
pathways to learning, 53, 58–59, 75
patience, 225
Pell Grants, 280
periodical indexes, 156
Perkins loans, 279
persistence, 225
personal:
    assistance, 10–11
    best, 311–312
    goals, 89
    mission, 89
    mission statement, 83–85, 104
    problems, 255
    relationships, 253–255
    shorthand, 173
    time, 70
    values, 82, 87, 272
personality:
    spectrum, 53, 60–63, 67, 75
    type, 60, 67–68
perspective, 19, 123–125, 154
persuasive writing, 177
Peterson's Guide to Jobs, 6
pharmacist, 35
physical:
    disability, 5
    science, 29, 37–38
    therapist, 35
physician, 35
    assistant, 35
physicist, 37
physiologist, 31
plagiarism, 184
plan of action, 123, 125
plan strategically, 125
planning, 178
plus loans, 280

positive effects, 120–121, 123
   of diversity, 239
positive self-perception, 70
power, 7
   of words and ideas, 167
PQ3R, 146–151, 153, 206
   reading technique, 139
   study technique, 137
prediction, 114
preferences, 73
prejudice, 240, 242–243
preparation, 227
   checklist, 182
pretest, 207–208
preview, 147–151
Preview-Question-Read-Recite-Review (PQ3R), 146–153
previewing devices, 148
prewriting, 180, 192
   strategies, 178–180
primary:
   audience, 177
   -level headings, 160
   sources, 139
principle to example, 115, 154
priorities, 82, 89–90, 97, 273
prioritize goals, 94
privacy, 295
problem solving, 113, 117, 119, 134, 264, 271
   plan, 120
   process, 119
procrastination, 82, 97–99
proof, 115
proofreading, 187
prosthetics technician, 35
purpose, 176, 181
   determines pace, 145
   for reading, 144, 161
   for writing, 176

qualifier, 210, 212
qualifying terms, 219
quantitative skills, 27
questioning, 133
questions, 147, 213
   based on the mind actions, 153
   in science, 313
quotation, 184

racial mixing, 240
racism, 242, 244, 247
radiation therapy technician, 36
radiologic technologist, 35

rating, 116
read:
   alone, 140
   critically, 151
   to evaluate critically, 145
   for pleasure, 145
   for practical application, 145
reader's guide, 157
reading, 137–155
   assignments, 139, 151
   audience, 177
   challenges, 164
   comprehension, 142
   critically, 155
   for comprehension, 163
   for critical evaluation, 161
   for general principles, 145
   material, 151
   note-taking and, 170–173
   overload, 138
   phase, 150
   place and time, 140
   for practical application, 163
   purpose, 144–146
   in science, 143–144
   skills, 138
   speed, 142
   strategies, 144, 145
   and studying, 142
readings in science courses, 143–144
real world perspective, 18, 43, 69, 91, 95, 124, 141, 245, 289, 303
reality resources, 267
reasoning:
   deductive, 115
   inductive, 115
reasons, 114
recall, 113, 120, 201
   improving, 201
recite, 150
recite, rehearse, and write, 202
recording information in class, 168
recreational therapist, 36
recycle, 307
reevaluate, 299
reference:
   librarian, 155
   materials, 139
   works, 155–156
refine, 119
reflective learners, 54, 64

registered nurse, 36
rehabilitation counselor, 36
rehearsing, 202
relating to others, 237
relationship:
   strategies, 253
   with words, 194
relationships, 253–255
   with people of different cultures, 247
relaxation, 209
research, 180, 230
   format, 187
   journal articles, 139
   reports, 187
   scientist, 224–225
   studies, 112
researching:
   grants and scholarships, 281
   a science education, 1
resources, available, 9–14, 32, 37–40, 42
   library, 155
respect for yourself, 246
respiratory therapist, 36
results, 190
resume, 277–278
retrieval stage, 201
returning:
   adult student, 90
   to school, 209
   to student life, 1
review, 150–151
   process, 144, 151
revising, 184
   and editing checklist, 187–188
Riley, Richard W., 5, 52
role models, 311

sabiduría, 74
sacrifici, 289
safe science, 223–228
safety eyeglasses, 226
salary/wage level, 275
Salvation Army, 307
sample and setting, 190
savings strategies, 286
scanning, 147
schedule, 91, 92
scholarships, 280–281
science, 225
   ability, 28
   books, 143
   courses, 144

discovering, 25–47
education, 2–4, 8–9
fields of study in, 29–42
knowledge and skills, 5
major, 44
and math skills, 2
as multidisciplinary, 296
-oriented publications, 143
readings, 138
researching, 1–20
student organizations, 14
texts, 138
today, 3, 294
women in, 42
scientific:
community, 144
discoveries, 27
discoveries and truth, 296
inquiry, 107–132
method, 112
research in the news, 48
self-:
awareness, 51–75
concept, 7
expression, 22, 49, 78, 104, 135, 164, 194, 220, 235, 264, 291, 318
image, 70
knowledge, 68
perception, 69, 73
talk, 142
senser learning style, 248–250, 262
sensory memory, 201
serendipity, 128
services offered, 13
set and achieve goals, 83
setting:
goals, 83, 88
priorities, 89
Seven Habits of Highly Effective People, The, 84, 311
sexist language, 187
shift your perspective, 123, 125
shorthand, 172–174
short-term:
and long-term goal setting, 126
goals, 81, 86–87, 92–93
memory, 201
scheduling, 101
similarity, 113, 154
skills, 27–28, 88, 271
analysis, 20
basic, 271

communication, 230, 271
lab work gives you, 230
marketable, 88
math, 2, 282
memory, 197, 200
observation, 28, 112, 232
quantitative, 27
science & math, 2
study, 58, 206–207
teamwork, 8, 230
technical, 231–232
skimming, 147, 151
slavery, 248
smoke signals, 17
social:
implications, 295
stratification, 172
Soloman, Barbara, 54
solving a problem, 117
soup kitchen, 305
specialized reference works, 156
speech pathologist, 36
spiritual life, 309
Stafford loan, 279, 280
standardized tests, 211
stereotypes, 240, 242
addressing, 242
stereotyping, 241
storage stage, 200
strategic job search plan, 277
strategic planning, 113, 126, 135
help you manage money, 282
strategic questions, 111
strategies:
can help you succeed on written tests, 209
for increasing your understanding of what you read, 142
to fight procrastination, 98
strategy, 125
is an essential skill, 126
strengths, 72–73
stress, 208
student:
handbooks, 14, 16
loans, 278–279
orientation, 9
population, 3
services, 13
study:
benefits, 62
groups, 152
materials, 207
plan, 208

reading materials, 146
schedule, 207
skills, 58, 206–207
techniques, 67
studying, 137, 147
a text page, 158
suà, 190
subjective questions, 211
substantiation, 115
subvocalization, 143
success, 300–305
and failure, 300
successful communication, 248
summarizing, 151
summary, 153
area, 172
surgery technician, 36

T-note system, 172
taking in, retaining, and demonstrating knowledge, 197
taking notes, 167–168
taking tests, 206
tape recorders, 204–205
teaching assistants, 12
team:
member, 258–259
success, 18
teamwork, 2, 17, 272
and diversity, 17
skills, 8, 230
technical:
skills, 231–232
writing, 143
technological growth, 297
technology, 31, 33, 51
test, 216
test anxiety, 206–209
and the returning adult student, 209
test:
directions, 210
items, 212
machine-scored, 210
mistakes, learning from, 215
preparation, 206–215
questions, types of, 211–215
standardized, 211
taking, 197, 207
type and material covered, 206
test-taking experts, 212
theoretical:
framework, 190
learners, 54, 65

thesis statement, 178, 181, 183, 193
think links, 150, 172–174, 178, 181, 214
thinker, 63, 262
   style, 249
thinking:
   critically, 109, 119, 185
   preferences, 63
   processes, 111–113, 116–117, 127
   skill, 112
Thinktrix, 113, 153
time, 89
   commitment, 273
   management, 81, 89–98
   management skills, 181
   management strategies, 96
to-do lists, 96, 102
topic, 176, 181
true/false questions, 213
truth, 296
tutoring, 28

unconstructive criticism, 256–257
understand and accept others, 238
understanding, 155, 243
   your values, 83
uniqueness, 19
U.S.:
   Bureau of Labor Statistics, 5
   population, 33
   students, 52

value, 116
   system, 82
values, 82–88, 100–101
   choosing and evaluating, 82
   and integrity, 272
   relate to goals, 83
verbal:
   information, 65
   learners, 54, 66
   -linguistic intelligence, 58
   -linguistic students, 67
   /linguistic learners, 65
veterinarian D.V.M./veterinarian technician, 36
violence, 254
virtual reality, 7
visual:
   aids, 202
   images, 204
   learners, 54
   -spatial intelligence, 58
   /spatial learners, 65
visualization, 172
vocabulary, 143
vocalization, 143
volunteering, 270, 305–306, 317

Web sites, 47, 130, 276
week-at-a-glance, 92
weekly and daily goals, 92–93
Weinstein, Bob, 9
wheel of thinking, 127
women:
   engineers, 42–43
   entering the work force, 5
   in science, 5
wordy phrases, 186
work:
   and school, 272
   -study program, 280–281
working while in school, 273
world competition, 52
World Wide Web, 157, 297
   science sites, 234
writer and the audience, 177
writing, 167, 175–187, 202, 228
   elements of effective, 176
   a first draft, 182
   good, in science, 175
   informative, 176
   an introduction, 183
   manuals, 184
   a mission statement, 84
   process, 168, 178
   process in science, 178
   purposes, 177
   an outline, 181
   schedule, 181
   skills, 8, 27, 175–176
   in a textbook, 148
   for typical readers, 177
   well, 167
written resources, 137
www.kaplan.com, 6

zoologist, 31
zoopharmocognosy, 32